Tea at the Cavendish

Ron Edge

DEDICATION

I wish to dedicate this book to my son Christopher, without whose assistance it would never have come about. His ability to get me to write it, and to see it through the publishing process were unbelievable. Perhaps his background helped, being born of Czech and English parents in Australia then growing up and getting his PhD at UVA in the US.

CONTENTS

INTRODUCTION

Some individuals are people-oriented, and some are thing-oriented. The ones that are people-oriented become ministers of religion, psychiatrists social workers and such. The others are engineers, car mechanics, brain surgeons etc.. You will find, reading this, that I am of the latter ilk. I am afraid, as a result, there is singularly little here about the way people react to one another, and their internal thoughts. There is a lot, however, on the way they behave. That's the way life is. If you wanted a psychological novel, seek elsewhere.

WHY BOTHER TO WRITE?

Have I had an interesting life or an exceedingly dull one? It has seemed interesting enough to me, but others might find it totally without redeeming features. Anyway, I feel imbued to commit my thoughts to paper for three reasons. Firstly my grandfathers, both of them, were dead by the time I was four, so I never got to know them, what they did and why. The only really fascinating events of my father's life – his experiences in World War I, including lying in a shell hole for two days, shot in the leg, with a dead comrade – he evidently found too traumatic to discuss with me – so, grandchildren, here I go, like it or not. In fact, the grandchildren may well find this too boring to read. Still, the manuscript may be useful to start a fire.

Secondly, scientists, and physicists in particular, are supposed to have uninteresting, boring lives. My feeling is that this is generally not so – the book by Watson "The Double Helix" gives some indication of this, and also Richard Feynman's book, "Surely You're Joking Mr. Feynman," has tended to put this myth to rest. Nevertheless, I feel driven by the muse (only question – which muse?). The third reason to write is – I feel like it, goodness knows why. I suppose I should start with "I was born at a very early age" – but I will spare you that. Nevertheless, at the age of ten I started a diary – it read as follows "Got up – had breakfast – went to school – had lunch – came home had tea – went to bed". After this had gone on for several weeks, I quit keeping a diary – except for odd occasions. This was at the beginning of WWII – to the adults, a catastrophe, but to us kids the most exciting and interesting thing that could have occurred.

I was always much younger than the average in my class, with the result I was bad at games, lacking strength, ability and agility. It was an all-male school, and I had no sisters – nor did any of my friends – and at Cambridge the ratio of men to women was eight to one, so the prospect of someone like me getting a date was zilch. In fact, it is somewhat of a surprise that I came out of this system reasonably normal socially, rather than being "gay" – but it took quite a while, and I am still perennially shy – I have to force myself to be social, (though few would think so!). I did not start to date until I went to Australia at the age of 25. There our accommodation in Canberra at University House, was co-ed – for me a very novel (and pleasant) experience.

1 CHILDHOOD

1.0 EARLY DAYS

Clip-clop went the clogs of the mill lassies on the stone flags – or more often, jingle-clip, jingle-clop – the clog irons, held on to the wooden sole by nails, like horse shoes, soon loosened, so that as the clog was raised, the iron would rattle. Clogs were ideal footwear to tend the cotton spinning "mules" which provided the primary employment in my home town. Samuel Crompton invented the "mule" there in 1750, and for nigh on two hundred years it was the support of the town. Yet it took less than thirty years for all this to vanish, and now one would be hard put to find a single mill in operation. Sic transit gloria mundi! (so vanishes earth's glory. A Latin tag gives tone!). There are no mill lassies left – "clogs and shawl" (shawl over the head) are long gone.

Lancashire folk had a saying: "Hear all say nowt, take all give nowt, and if ever tha does owt for owt (anything for anybody), do it for thisen (yourself)" – a very negative attitude, but one adhered to by the prosperous mill owners! As the son of a white collar worker, I was destined for Public (private) School, rather than a life in the mills, so inevitably we lived in a semi-detached house in the suburbs – all "mod cons" (modern conveniences - an indoor toilet). The semi is an efficient plan, in that the two adjacent accommodations have a shared wall for the chimneys, thus avoiding heat loss to the outside. However, the lead plumbing was on the outer wall, so that it inevitably froze each year or two. It didn't start leaking until the thaw came, and the plumber would be called to "wipe" the point where it leaked to seal it with molten lead. One never drank hot water from the tap – it contained dissolved lead salts. Even today, when this hazard is long gone, I forbear to drink water direct from the hot tap. The hot water came from a small copper box behind by the coal fire, connected by two pipes to the "cistern" upstairs. Convection carried the lighter, hot water from the top of the box up to the cistern, far above in the bathroom, and the colder, denser water from the bottom of the cistern returned to the box via the other pipe. You see, it is in my blood to give such a physical explanation! The only time I saw my mother nearly in tears was when my dad did not want to spend the money to install an electric water heater to replace this cumbersome and inefficient system. Then, I realized – we were poor!

In winter, life centered around the coal fire. Setting and starting the fire, then getting it going with the "blower" (a metal sheet which you held in front of the fireplace opening to provide a draught) was a tedious business. Sitting around the fire was very cozy however.

The washing was was boiled over the gas ring – mother never had a washing machine the whole of her life.

In winter we wore wool underwear. I still itch at the thought of the wool vest and underpants – but they did keep you warm. The acrid smell of drying wool is still in my nostrils. Wool was also used for swim togs, which covered the whole body, even for boys. After the bathing suit got wet, it became heavy and most unpleasant. But then, the incentive to swim was very weak, the temperature of sea water being about freezing most of the year. Our holidays were restricted to Blackpool and Colwyn Bay during the war. I did not go sea bathing the last ten years of my life in England. Ah, the balmy summer sea water here in South Carolina!

Dress as a child was so different then. Boys up to age twelve – puberty – wore short pants, of gray flannel, white shirt, pullover and tie, and generally a jacket – also a cap which I hated, with a school badge on it. Also braces – what are referred to in America as "suspenders". I did not replace braces with a belt until well into my teens. Looking at past photographs makes me wonder – who determined this weird dress code?

1.01 THE VICAR

My parents practically never went to church. Nevertheless, in true British tradition, they were considered Church of England. So now and then, the vicar would visit. He was a pleasant, laid-back man, easy to talk to. My mother would entertain him in the front room, which smelt of moth balls and mold, because it was so rarely used. Once, for some reason, my mother wanted to entertain him in the living room, where I had all the chairs arranged in line to play with to simulate a train. In five minutes, we had it all straightened up as he walked down the path, and he had a cup of strong tea with us. The only time I recall my parents going to church was when I was confirmed. I think my father did not want to contemplate the after-life – or anything connected with it.

1.02 BIRTH

I don't remember much about my birth! The thing was, they had a craze for maternity homes about that time, and so my mother went to the nearest – the Haslam Maternity Home. This was, in fact, a very large house, originally the home of a mill owner. It must have been an impressive edifice in the nineteenth century, a rather cold building all built of stone, and it had aged pretty well, though suffering from the arthritis of old age in the plumbing. It was however rather grim, and would have made a suitable setting for a murder mystery. Luckily, all went well, and the doctor (Dr. Wright – he remained my doctor the whole time I lived in England) helped me into this world. What was so interesting is that the maternity home later became a geriatric residence, and my mother finally died there. I gather it had changed little in between. Life does some strange things to us.

1.03 BUSES AND TRAMS

I remember catching the double decker bus on a cold, rainy winter afternoon – the rancid smell of wet wool and the mill lassies shouting at one another. Why? Because they were returning home from the cotton spinning mill, where they had to shout over the noise of the machinery – and, even though they did not need to on the bus or tram (trolley car), habit made them yell. A famous singer of the 1930s ("the Biggest Aspidistra in the World", "Walter, Walter lead me to the altar"), Gracie Fields, developed her powerful voice at the mills. Another thing: the mill workers were always small. I was standing outside a mill onetime when the mill workers were leaving. Of this large crowd, I could see over the heads of all of them and I am not tall (5 ft. 8 in).

The trams were delightful for kids to ride. All were double-deckers, but some had only four wheels, and others had two sets of "bogeys" (trucks) each with four wheels pivoting together. The four wheel trams had to have a narrow wheel base to negotiate comers, and were inherently unstable, with the result they would jump the rails (tracks) periodically, rarely hurting anyone however. If you sat on top at the front, you were subject to at least three independent oscillations – up and down, side to side, and a curious undulating motion as well. It was better than being on a ship at sea! and coming down one particular hill, the tram moved faster and faster, and there was always the hope it would hop off the tracks as it rounded the curve at the foot of the grade (as had happened once or twice before). The bodywork of the car was largely wood, nicely painted and varnished. Trams were double ended, with a spiral staircase, brakes and an accelerator pedal at each end – but the driver (who always stood while driving) took the accelerator pedal with him when he changed ends. The seat backs could be pushed over or reversed, to allow the seat to face in either direction. They were noisy, but not smelly. Even today, given the chance, I ride in front on top of a double decker.

I believe the Johannesburg trolley buses still run regular double-decker routes – but when I was there, only blacks could ride them! Open top double-decker trams still ply the sea front in Blackpool. Double-decker buses are interesting too. To reduce their height to pass under bridges, the Ribble company buses were made with the upper deck very low – with a deeper passage down one side of the bus along which one could walk, then slide along the seat – not easy! My maternal grandfather (grandpa Davies) liked the "toast-rack" trams. These were single-deckers completely open on both sides, with seats running fully across. One could get in anywhere along each side, making the vehicle look like a toast rack. The conductor had to swing along the side like a monkey to collect fares. An eccentric uncle, in the days of the horse trams in Douglas, Isle of Man, would get off and, irritated at the slow pace of the vehicle, walk beside it while his wife sat inside.

1.04 JUMPING OFF THE BUS

I rode home on the bus often, and to shorten the walk home, I would jump off as the bus slowed down rounding the corner near home – and I have fallen, grazing my knee when I missed. The old buses had a wide exit at the back. You held on to a pole linking the floor the ceiling, put your left foot out and let go. I tried it once on a left hand drive bus in America – but it didn't work – I was too used to the right hand drive.

1.05 MEMORIES OF SIGHTS AND SOUNDS

The warm glow of the gas lights down the avenue, the horse and cart that brought the milk, which the farmer ladled into my mother's jug with a metal dipper measure. The horse, at a whistle signal, would walk on to the next house, saving the farmer steps. I believe the milk gave me TB – but it never bothered me.

The gas lights were quite interesting. In an earlier era, they were lit by the lamplighter at night and turned off in the morning. The up-to-date system we had was worked by a pressure pulse of gas in the evening applied at the gas works, which traveled up the gas line, and rotated a mechanism in the lamp which allowed gas into the burner which was kept lit by a small pilot light, that was on permanently (unless blown out by a wind gust). A similar pulse in the morning turned it off. You could watch this pulse traveling along a straight road, the lamps turning on one after the other. This worked quite well, except that once in a while the pulse failed. The result was that the lamp turned itself off at night, and on in the daytime. So, an employee of the gas company had to come round now and then and set this straight.

1.06 PARENTS AND GRANDPARENTS
1.06.1 Father

My father was born in poor circumstances and never finished high school. He worked in a warehouse, where his hand got caught in the lift (elevator) door. The result was that, even years later, his left hand looked like a fried egg, sunny side up.

When my father died at 79, my mother gave me the letters he had written to my grandmother in the first world war, for which he volunteered illegally at the age of fifteen. I asked my aunt one time why anyone would be so foolish as to do this. She told me that a Scottish regiment came marching through town recruiting. They came in kilts, with bagpipes playing, flags flying and so forth. How, said my aunt, could any impressionable fifteen year old youngster in a dreary mill town resist such a call?

He and a mate were running dispatches, got shot at and ended up in a shell hole. My father spent a couple of days in that hole, shot in the leg, with his dead companion, before being rescued. He never spoke to me about it. I have always wondered why. Was it too traumatic? Was it that he thought I would not be interested? I would have loved to know about his experiences – but now it is too late. I would hate my kids to feel the same way – which is one of the reasons for these memories.

He had been in two Scottish regiments – the Argyle and Sutherland Highlanders, and the Camerons – which is why, when I took up Scottish dancing, I got a Cameron kilt. However, the dress kilt, which I bought, is much more colorful than their fighting kilt.

The Scots celebrate Hogmanay (New Year), not Christmas, so my father as a sassenach (non Scot) was put on guard duty New Year's Eve, with the job of putting all the drunken Scots in the guard house as they rolled in. He was only sixteen at the time.

Dad was a very simple man – but he got on with everyone. I recall going for walks in the evening with him when I was under much stress taking exams. Who knows what we discussed, but I did get to sleep afterwards. Being an only child, I got much attention, but equally, much was demanded of me – although I never realized this at the time. I had to work hard evenings – but they put a little fire in my bedroom fireplace where I did my homework, even during the war.

Returning from WWI, he lacked ambition – he worked first at a Bessemer steel converter until it went broke, then as a laborer in a leather tannery – in the tan-yard – terrible smell. I went there once.

He took my mother's fancy – she was a secretary, and daughter of the manager, who, surprisingly, did not discourage her – her Welsh blood, the Celtic temperament, probably meant she would go her own way anyway – strong minded – or stubborn, whichever. They "kept company" – dated – for quite a while, because he was supporting his parents, who had no money. My mother's father's wedding present was the down payment on the house, where they lived the rest of their lives. They were happily married – it gave me a warped view of marriage, which, experience has taught me, is rarely like that – though I am slowly recovering from that point of view

1.06.2 MOTHER

My mother did not have the leaning towards learning which my aunt had. After finishing high school – (she went to the local grammar school), she attended a secretarial school. She learned Pitman shorthand, (which later I tried, and failed, to master, but odd bits are still useful in taking notes) and went to work at Walker's Tannery. One summer, she performed in a pierrot troupe, and I have a photo of her with a tall conical hat with bobbles on it, and a floppy dress also with large bobbles. Such troupes were popular in England during the summer, performing daily, and traveling about locally. A typical joke – "Are you a piece of toast?" "Why?" "Because I'm a fried egg and I want to sit down" Ugh!

She led a bright social life, but fell in love with my father, who was a lowly laborer in the tan-yard. They courted and married and lived a very normal life. Luckily, my mother looked after the finances – she was much better at keeping accounts than my father. She never had a refrigerator or a washing machine – so she went out shopping every day, and met all her friends and neighbors – and when I came home, I went too, so she could show me off. I think this walk to the shops each day kept my mother healthy. My parents never had a car, although my father had a Rudge motorbike when courting my mother. The tale went that the bike would not start once stopped, and one time they were going on vacation to Clitheroe. Losing their way, my father would yell at a surprised passerby "Clitheroe, Clitheroe?" pointing the direction but without stopping, much to their surprise.

My mother was a very loving person, but she and my wife Margaret did not get on – and I was in the middle. The Scots have a saying "beware the redding straik". If you try separating two people fighting, you will get struck yourself.

1.06.3 GRANDPARENTS

I recall little of my mother's father, Walter Davies, and my father's father grandpa Edge, was already dead, about the time I was born. My mother remembered her grandmother Grandma Redford – Anne, as a little old lady who sat in a corner smoking a pipe. My mother's mother I remember quite well, as a small but evidently feisty individual. She lived on her own after her husband went, until the day she died, whereas my father's mother, Grandma

Edge, was a placid relaxed individual, who used to give me, as a four or five year old, my favorite candy, "mint imperials" – last time I saw this kind of sweet was when I visited South Africa. It is as well that she was placid, for her husband left her penniless, and she had to live with us in the spare room for a few years. As she was dying, my mother told me she thought she was singing "bringing in the cheese" – it turned out to be a well-known harvest hymn (though not to my parents) "bringing in the sheaves"! I also recall the collection hymn children sang in Sunday school "Hear the pennies dropping, listen as they fall. Everyone for Jesus, he shall have them all". My great-grandfather on my mother's father's side was from Shopshire – theWelsh marches – probably a small village called Dovaston, since that is my middle name. It was also the name of one of my grandfather's sons who died in infancy. Great-grandpa was a coachman, and died as the result of a horse kick. I know little of my father's relatives. They lived in Breitmet, at the other end of town. My mother had some interesting and eccentric relatives. There was auntie Susie, whose husband had a player organ – a rare device which you pedaled like a player piano, the music punched on a paper roll.

Whereas my mother's father was manager of the "Rose Hill Tannery", (in spite of the name, it was a smelly, dirty environment – but "where there's muck there's money" is a famous Lancashire saying) where my father met my mother, my father's parents lived from day to day – if they had money, they would buy a chicken to eat – a great luxury then. My mother's people planned carefully for the future, which was just as well since grandpa Davies died relatively early, of a thyroid condition called Grave's disease. Anyway, my parents had to support my father's parents for quite a while.

1.06.4 ELEMENTARY SCHOOL

I started school at the age of four. In America, generally kindergarten comes first, but in England, at that time, you jumped right in, so this was not unusual. The first day I was astounded because the boy next to me wet his pants – a thing I would never do! It was in Miss Peregrine's class, a rather small woman, appropriate to the first grade class. There were two senior class teachers. One, Miss Dearden was rather deaf – and a little old! Luckily, I did not have her.

Instead, when I was nine, I got Miss Fletcher. She was a large, jolly, woman who made us take intelligence tests on a regular basis – this improved our intelligence, or more particularly, our ability to take these tests, which were essential to getting into a good high school.

I went to school on my own – about one and a half miles, walking or taking the bus in wet weather. My memories of grade school are vague – I recall "Empire day" – the headmaster would call us out to the school quadrangle, and lecture us on the benefits of the British Empire – still going strong then – and we would wave little union jack flags. Summer vacation we would go to Morcambe, Blackpool, Fleetwood, Colwyn Bay, or Llandudno – and I would sail my toy sailboat on a shallow lake.

One summer we went to the Isle of Man – dad and I were incredibly sick on the sea passage over – not so my mother. We never went there again. I recall as we got off the boat from Fleetwood to Douglas on the Isle of Man, the world seemed to be spinning around. The old joke was, the steward says "you can't be sick here!" My father said "Oh can't I?" (Was).

1.06.5 MARIONETTES

Every summer, a marionette theater opened in a shed near the beach in Fleetwood. It was run by one man (sometimes two) and for the most part ran pantomime material such as "Aladdin" or "Puss in Boots". I loved it, and as a result, built my own. I was terribly taken by special effects – so I had a skeleton which dissected itself and a troupe of chorus girls, whose right legs were all worked by the same string, so their high kicks were perfect. The Punch and Judy shows on the beach were good too. I always wanted to run one, with a Toby dog who walked around with a ruff round his neck and a little conical hat, carrying a little bag held by a wooden handle in his mouth. What child could resist putting coins in his bag when he wagged his tail? We can't do that today – I suppose it all vanished with modern civilization after WWII.

1.07 HIGH SCHOOL

At age ten it was imperative to do well in the city examinations to enter a good high school. It was at this point that the sheep were separated from the goats, and since my parents could not afford to pay my fees for Bolton School, I had to get a scholarship. The exam is now called the "eleven plus", but I was only ten when I took it. Intelligence tests had just become the thing, so as mentioned above our teacher had us take one or two a week – amazingly, you can really improve if you do this – though all the books say you shouldn't. Anyway, I got the best scholarship to the best school – so not to grumble! It so happened that this was right at the outbreak of war – 1939 – so the first day at this somewhat overpowering school built in mock Tudor architecture (it looks like Hampton Court Palace in red sandstone) was chaotic. I went to the wrong entrance (they were finishing the bomb shelters) and was directed by the "sergeant" – the sort of concierge. The sergeant (he really was an army sergeant – retired) was in charge of making everything run smoothly. Erected by the first Lord Leverhulme from the profits of Sunlight Soap, the original school had a history dating back to 1540. Nothing remained of the old building, the new one being a mile or two from the town center on Chorley "New" Road (at least several hundred years old!). I cycled to school every day. The bike was part of the reward for winning the scholarship. It was second hand, but a racing bike with rams horns handlebars, as opposed to the more conventional "sit up and beg" bars.

The war had just started when I entered high school, and they had just completed the air raid shelters against the Jerrie's bombs. The first year, there were daylight raids by the German bombers, so we were all sent down to the shelters, which were dank and murky – and exciting. It is difficult to believe, but the break in the monotony of school life came as quite a pleasant relief – and we would play cards. No bombs fell even remotely near us – but a fighter plane flew into the roof of a house opposite the school. The tail of the fighter stuck out of the roof for quite a while. I never heard the whole story of this, but my feeling is that it was an old school boy pilot showing off – when he lost control. The nearest bomb fell about a mile from my house – probably jettisoned by a German bomber which got lost. I cycled over to see the crater, which was quite spectacular. It had filled with water when I got there, but was many tens of feet in diameter.

When I first went into our chemistry lab I was astounded to find the walls and ceiling looked as though they were suffering from some obscure disease – they were mottled black and yellow. It turned out they had been painted quite recently, but white lead was still being used as the pigment, and the sulfurous fumes (H2S) used a lot in chemistry had turned the paint black. It was very depressing, but during the war no repainting could be done.

1.07.1 DRESS

As a small child, my clothes consisted of short grey pants and a white shirt, together with a short, or maybe long sleeved gray woolen pullover . The worst part was woolen underpants and vest (undershirt). They itched abominably, but were necessary to keep you warm. On a wet Monday, they were washed and hung on the 'maiden' (a wooden frame) in front of the fire to dry – and the acrid smell of drying wool is something none can forget. On entering high school, our dress was similar, but I wore a blue blazer, with an emblem on the pocket, and a school cap, (also with emblem), which I hated. The big deal was that on reaching puberty, we were allowed to wear long pants – also gray flannel – and of course, a school tie. This was a piece of white cloth with horizontal black stripes – very unromantic. However, as we got older, we acquired a more dramatic tie, diagonal stripes of various colors, meaningful in that each school had its own.

1.07.2 MASTERS – AND MISTRESSES

My memory of the masters and mistresses (note – in England "mistress" is the term for a female teacher) at school is somewhat vague – apart from two or three. My earliest memory of high school is Madam Saxelby. She did not run a brothel, but taught French, which is why she always held the title "madam". She was an extremely lively person, and her first accomplishment was to provide us with French names. Mine was Egu – from accent egu. As a result of this, I held the nickname "eggy" for the rest of my school career – much to my dismay. Luckily it did not persist to Cambridge.

Then there were Mutt and Jeff. The two Jary brothers taught chemistry, and since one was Geoffrey, they had to have the names of two comic characters in the funny papers. Mutt had been an industrial chemist for many years. He worked for an outfit which, among other things, extracted silver from lead. In early times it was uneconomic to extract the silver from lead, both often occurring together in the ore. However, modern techniques make this highly profitable. Jary would put on his best suite, go to the vicarage of an old church, and start a spiel to the vicar that he had noticed the roof of the church was in sad need of repair. Out of the goodness of his heart, he was prepared to give the church a brand new lead roof at no cost. They would then take down the roof, restoring it after extracting the silver, making themselves a tidy profit in the process. Jary was fond of pointing out that he had a pachydermatous skin, years of dealing with toxic chemicals having made his hands impervious to most things – including heat.

Mr. Blair taught us physics. He was relatively short, with hair that stuck up a bit. I always enjoyed experimental physics – still do – and we had a wide range of equipment. He enjoyed art too, and I have one of his engravings of the entry gate of the school. Since this was completely symmetrical, it was identical to its mirror image. Hence, you did not have to worry about the fact that the print and the engraved plate were mirror images. I never found out if he did this deliberately.

Miss Waterhouse taught biology, and had some sort of a relationship with the biology master, Sam Loring. I believe they may eventually have got married. This was the cause of much speculation among the boys.

1.08 RELIGION

What should I believe? I was christened a Methodist – because my grandfather was manager of a tannery, and the tannery was run by a Methodist family. Today, the chapel is a shopping center. Later, I became Church of England – since most of my friends were C of E – and was confirmed (the only time I saw my parents in church was at my confirmation). I was married by the Coadjutor Bishop of Canberra in the Anglican church in Australia, at Canberra. Even the future prime minister (Bob Hawke) and the American Ambassador attended – no other amusements that day!

Religion was an interesting problem at school. Having an established church, most English schools (including ours) had a Church of England service at the start of each day, those of the Jewish faith remaining outside.

We also had a religious studies class once a week led by a minister of the Church of England. He must have been exempted from armed service for some reason, which made him an odd bird to start with. We spent most of our time arguing with him, which he enjoyed, and made us think, something which students loath, but I shall never forget the benefit I got from this. He himself had doubts, and said he was a "Christian agnostic", a description which fits me too. Though, as a Unitarian, many would throw me out of the Christian church, the fundamental ideals of Christianity, love thy neighbor etc., are still paramount.

At my mother's behest, we had our first child, Christopher, baptized in the Church of England at the small village of Belmont, on our way to America from Australia. Her best friend was wife of the minister there. This caused some problems when my son Christopher later converted to Catholicism. You can only be baptized once, and the Catholics demanded evidence of this. So Chris wrote to the Belmont minister for a certificate of baptism – but he didn't say why. The minister sent it back with the remark "for whatever reason you need it – here it is!"

1.09 AUNTIE ADA AND BLACKPOOL

As a child, I well remember going on vacation to my Great Aunt Ada Sutcliffe's place She ran a boarding house in the holiday resort of Blackpool. Her husband had died early, and her sole support was to rent rooms (bed and breakfast) to holiday makers in the summer. At that time, cotton ran Lancashire, and each town (Bolton spun cotton, Rochdale wove it) took a one week holiday when the whole town shut down – but everything, shops, police, everything. Then they all went to Blackpool, and saw the same people they would have seen at home, but on the beach. So my great aunt would rent out rooms, usually for the week, and she, with possibly one assistant, worked like crazy. Everyone had bacon and eggs for breakfast, so she had the largest frying pan I have ever seen – it must have held easily twenty fried eggs and bacon, on the gas stove.

During World War II her house was occupied by the air force, and had future pilots billeted on her. Many later died in dog fights, or bomber crashes. It was tragic. One airman liked my aunt's place so much, he and his family took it over and ran it after the war, in return for looking after my aunt, who was getting old. She moved to London (where they had come from) to help take care of his kids, and as a young teenager I visited her there. She was as deaf as a post, and one time I took a tube train trip with her (she was then about 80) I noticed she bought no ticket, and at the end of the line, the ticket collector asked for hers. She totally did not understand him (so it would seem!), and after about ten minutes of futile explanation, while a long queue of irate passengers built up, he let her (and me) through. She never said anything, but I always wondered whether this was a regular occurrence.

Blackpool in the thirties was a happy place. People did not have enough money to go abroad, nor had the idea even occurred to Lancashire mill workers, who were very suspicious of foreigners, although very friendly once you got to know them. The bracing sea air ("breathe deeply" my father would say, "it's the iodine that does you good"), walks along the front or promenade, the "South Shore" fun fair and donkey rides made for a fun holiday. Swimming, or paddling, in water invariably as cold as ice – my father would enter the water, swim around and come out looking blue. I avoided getting a "chill on the liver" if possible. This was a common complaint – I have since discovered that a chill on the liver is a purely English disease – they are the only people who suffer from it. In my case, it turned out that what I thought was a chill on the liver was migraine, a totally different complaint. I happened accidentally to mention the symptoms of my chill on the liver to my doctor ("the quack" we always called him). "Oh, you have migraine" he said, and prescribed fiorinal tablets for it. You had to take the pills ahead of time if you felt an attack coming on. If it was not migraine, they had no effect, but if you had migraine, it not only cured it, but gave you a terrific high.

Auntie Ada had lived in Bolton during WWI. The "Bosch" (Germans) sent over Zeppelins to bomb England. These were primitive hydrogen filled cigar shaped dirigibles. The crew would lug the bombs over the side to drop them. For some unfathomable reason (most probably they were lost) they dropped bombs on my home town, actually killing one or two people. They came over at night. My Aunt Ada was so deaf that she slept through the whole thing, and was terribly disappointed and annoyed she had missed the excitement.

1.10 THROWN OUT OF COURT

I took a deep interest in the law at about age ten. So, I decided to go down to the local court to see how justice was administered. The court was housed in a building close to the town hall, and there was a sign outside showing where the public gallery was. I ascended the stairs and sat down in the first row overlooking the court room. I was the only spectator. The first case was about a burglar who had robbed a child's money box. Curious. The judge examined the money box carefully to see how the thief had broken in. At about the same time he noticed me in the gallery. Peering at me through his spectacles, he said "How old are you?" "Eleven, sir" I said. "Well" he said "You are not supposed to be in here unless you are at least sixteen. You had better leave." I got up, very embarrassed, and left. The next time I was in court was on a capital jury in America.

1.10.1 MUSTARD, PEPPER.

Viscount Leverhulme, the soap baron who essentially re-founded my school, was the son of a chemist (pharmacist) who lived in Bolton before the turn of the century. Small children were sent to pick up items from this chemist, who had a rather curious sense of humor. He would say, "What do you want — mustard, pepper, starch, or stone, glue, resin, pitch, or tar, rice barley, sago tapioca, blacking, blacklead, vermicelli, furniture paste, brick dust, jollap or rhubarb?" Chemists today would be totally unfamiliar with these items, which were easily available in the late eighteen hundreds. After this barrage, the kid had to go home because he forgot what he came for! The list was taught me by my father — goodness knows where he got it from, or why he remembered it — and I have no idea why I remember it too!

1.11 AUNTIE GLADYS

My Auntie Gladys was my favorite — and only — aunt. She was about three years older than my mother, and whereas my mother was fun loving — she liked to go out dancing and was quite satisfied with being a secretary when she worked, Auntie was more academically inclined, and went to college at Bingley, in Yorkshire, to train as a teacher. So she taught school for a number of years, which proved useful when my uncle lost his job because the solicitor for whom he worked went to jail — some said on account of the incompetence of my uncle — however, I used to have nice discussions with my aunt on many subjects. My mother and I would go down to Leicester by bus to visit my aunt after she moved south, and I would go around the museums there. I recall seeing the pants of the fattest man in the world in one museum.

It is strange, my cousin David was good at sports — which I was not — so he got on very well with my Dad, (who was very athletically inclined) after I left. Dad played tennis until his forties, then switched to golf. Lawn bowls was more in my line — you could spend time deciding what to do.

At college they had a set of bowls each shaped like the frustum of a cone — these were similar to the bowls Drake used on Plymouth Hoe while awaiting the Armada — they had a trajectory on the green like a walking stick — almost straight, then a sharp curve like the handle. The only problem was we used the college fellows' garden, and the bowls had a habit of surmounting the elevated path around the garden and plunging into the river Cam.

To return to my aunt, I used to stay overnight at her place in Leicester (or Kirby Muxloe to be accurate), which had a nice fifteenth century castle. Kirby

Muxloe was about half way going to and from Cambridge by motor cycle (on my BSA bantam 125 cc) and the stop made a pleasant change for me. Once we visited Ashby de la Zouch castle, built by the same baron who built that at Kirby Muxloe, The castle keeper showed us his pet bat, which he kept in his pocket.

1.12 CHANGE RINGING

At the age of ten or twelve, I read a book about the English sport, (or "exercise" as it is called over there) of change ringing on church bells, and went so far as to tune eight empty pop bottles by filling them with water to the right pitch, so that I could practice ringing changes by tapping on them. However, not until I was at university did my interest become strong enough to induce my father to get one of his friends to start me. I am sure I had read Dorothy Sayer's book, "the nine tailors", which is an excellent introduction (she was a member of the Oxford University Society, whereas, of course, I belonged to the Cambridge University Guild).

The friend was Peter Crook, a well-known Bolton policeman, whose son was a detective inspector. Both were big men – with a name like "Crook", as policemen they probably needed to be, but Peter was a gentle, gregarious soul. I recall going over to Deane Church – an old building, to watch the ringers for the first time and being quite fascinated with them, seeing the ropes fly up and down, each one through its own hole in the ceiling. I was "hooked" from the word go. I walked from home to the church, crossing the "middle brook", where I used to catch minnows and sticklebacks with a little net, and through Deane Clough. A "clough" is a little valley – I discovered in South Africa, the same word "Kloof" is used. Peter's home church was Trinity, downtown by the railway station, and there he undertook to teach me change ringing. The church is now defunct, and the bells have been removed – a ringer sent me a picture.

The ringers formed a closely knit, friendly group, and it has been my experience throughout the world, whether it be the US, Britain, Africa, Canada, or Australia, that a change ringer is immediately accepted into the community of other ringers. Of course, not all countries have English style church bells. English change ringing differs from all others in that the bell travels a whole circle at each ding (or dong!). It is mounted on a horizontal beam or "head stock", which rotates on bearings at each end, driven by a large wheel round which the rope winds, and for which it is the axle. The bell starts mouth down, but in ringing, successive pulls swing it higher and higher, until it is balanced mouth up, where it comes to rest because a wooden "stay" attached to the head stock, comes up against a "slider", which stops it going completely over. It can then be left in this semi-stable position.

The rope is attached to the wheel about three quarters of the way around, so it unwinds as the wheel rotates, until this point is opposite the hole in the ceiling (through which the rope passes). Then it is drawn up again a short distance by the wheel until the full circle is completed. This means, as you pull the rope with the bell mouth up, the bell will come off the slider, and the rope will unwind – but before the full circle is completed, it rises, and the ringer catches the rope and pulls letting it reach the point of balance again. Pulling on he rope at this time will cause the bell to swing in the opposite direction. The end of the rope, held when the most rope is wound around the wheel, is called the tail, and the pull given here is backstroke. The rope then falls in front of you, as the wheel rotates, and may flap about frighteningly unless you have give it a good straight pull. Then, up it goes, but only a short distance, and must be caught on a tufted portion called the "salley", at "handstroke". With training, one can control the bell, and pull it off at the right time to strike immediately after the previous bell, not too early and not too late – "even" striking, which puts the heart of a bell ringer at ease.

Change ringing is not like any other sport – it is not solely intellectual, like chess, nor is it mostly brute force, like Rugby, but it requires both skill and energy, together with a considerable amount of mental concentration. The culmination of all this effort is to ring a peal. There are 5040 different combinations of seven different numbers and a peal on seven bells is the successful ringing of all of these, which takes about three hours.

1.12.1 DEATH OF GEORGE VI AND CORONATION

Upon going up to Cambridge in 1947, I became a member of the University Guild.

We rang a fully muffled peal for the death of George VI. The bells had their clappers covered with leather, which produced a dull, heavy sound, suitable for a funeral. We rang a happy peal for Elizabeth's coronation. One of our ringers was a peer – a baronet I believe. Anyway, he attended the coronation in Westminster abbey. Because of the crowds, the peers had to arrive very early, and brought sandwiches for lunch. They were all in their robes and coronets, which they wore. Some had thoughtfully stuffed their sandwiches in their coronets since they had nowhere else to put them. The ceremony went well, until the point where the queen was crowned. The peers were then supposed to raise their coronets and say "God save the Queen". Many had forgotten to remove their sandwiches, which proceeded to fall out of the coronets, causing some consternation.

For normal funerals it was usual to have the bells half muffled. This meant one side of the clapper was covered with leather, but not the other. This gives a delightful echo effect – first the changes loud, and then soft. We rang in whole pulls, the same for both hand and backstroke.

There are two famous bell ringing societies in England (and the world) The Royal Society of Cumberland Youths (most "youths" are over 50!) founded in, 1747 and the Ancient Society of College Youths, founded in 1637, both in London. I belong to the Cumberlands.

1.12.2 EARLY TELEVISION

Around 1950 or so the Cambridge University Guild of Change Ringers held a tour in Devon. I got a ride with a member of the guild who was an engineer with the BBC. In fact, he was the original engineer whom the BBC television service hired. His tales of the early days of television were fascinating. Service began just before WWII at Alexandra Palace, just north of London, which still (2002) has the broadcast antenna on top of its tower, clearly visible from the railway (GNER). The BBC had not decided which broadcast system to use – two were preeminent, one run by EMI (electrical and musical instruments) and the other by Baird (John Logie Baird, who was one of the early inventors of television)

As a result the news reader (always dressed in white tie and tails) was televised by both systems simultaneously. Since neither wanted to reveal its secrets to the other, they were furiously concealing their equipment, at the same time that they were employing the two cameras to view the newscaster. Both signals were then broadcast. Ultimately, EMI won out, but then the war came and everything closed down. The picture had only 440 lines, making it blurry. It shows the disadvantage of being first. The changeover to 560 lines was very tedious.

We drove on down to Devon, stopping off to see Stonehenge, which at that time was most impressive, standing completely alone, no fences or buildings, as now. I recall blowing a cylinder head gasket in Paul Williamson's car in the middle of Dartmoor. Somehow we staggered on to the nearest town, with the car blowing steam in the back, like a locomotive. One of the ringers managed to fix it. We rang at Widecombe in the Moor – home of Tom Pierce and his gray mare in the song.

The younger members of the Guild used to hold a cycling-ringing tour which involved staying at youth hostels. In those days, you had to arrive at the Youth Hostel on foot, or possibly by bike, to be accepted. I had been running on the nuclear accelerator doing research at the Cavendish, which had overshot time wise, with the result I could not bicycle down in time. However, I could, and did ride my 125 cc motorbike down to Kent, where the ringing was, and participated. This meant I would arrive at the Youth Hostel on my BSA Bantam motor bike, stuff it in a hedge, and walk up to the door nonchalantly. The Kent tour was most enjoyable, because otherwise I had few friends my age.

Since each bell in the tower rings once at every turnover, one cannot ring tunes, as with a carillon, where one bell may strike a few times before the next. As a consequence, the English developed a method of ringing called "changes". The first rule is that, at each pull, all the bells strike sequentially, but the sequence changes each time

It takes about three hours to ring a "peal" on seven bells – factorial seven, 7x6x5x4x3x2xl, each change being different giving 5040 changes. This can disturb the local neighborhood. However, at least in England, practice night is sacrosanct. If a band has been practicing every Tuesday night for the last two hundred years or so, the law says you can go on doing it – the neighbors knew about this when they moved in. However, in America it is different. With the new bells at Washington cathedral, a nearby retired supreme court justice threatened a law suite if they continued.

I recall ringing a peal in Lancashire and on finishing we noted a collection of burly Rugby players waiting for us outside, obviously irate at having their football game disturbed. Under these circumstances, the usual thing is to outwait the footballers. Eventually, they will go away. Unfortunately, one of our group had to catch a train. So he slipped out the door, saying to the rugby players – "I've stopped them!" We stayed put inside.

1.12.3 CAMBRIDGE UNIVERSITY GUILD

Upon going up to Cambridge, I naturally became a member of the Cambridge University Guild of Change Ringers, and ultimately graduated to becoming secretary. I had many interesting experiences with the guild. All the members were "characters", the like of which I have rarely seen elsewhere. We bought an old London taxi cab to travel about, one where the driver must sit outside, in all weathers, wrapped in umpteen overcoats. The ringing band sat comfortably inside, and since there were a couple of jump seats, it was possible for four people to sit opposite, and ring a peal in hand (using hand-bells) as we traveled along.

At that time, there were two Guild Weeks, touring the country ringing at different churches. Hand-bell ringing is used to complement the tower bells. Again, changes, not tunes are rung. In America, the hand-bells are used mostly for tune ringing, as a result of a peculiar circumstance. The tale goes that Phineas T. Barnum saw these English hand-bell change ringers on a visit to Great Britain, and was quite taken with them, with the idea of putting them in his circus. However, the British ringers were wearing dull British clothes, so the first thing he did was dress them in colorful outfits he happened to have to hand and call them the "Swiss Bell-Ringers". The second thing was have them ring tunes, instead of changes – and so it is to this day, though most people don't know why.

If we occupied a railway carriage it was nice to have it to ourselves. Hand-bell ringing proved a solution. We would use our thumbs to represent the bells. Four of us would sit and ring "bob major", a method of ringing, using our thumbs to represent bells – thumbs up was hand-stroke, thumbs down backstroke. After the other passengers had watched us putting our thumbs up or down interspersed by someone calling "bob" and "single," the carriage would clear at the next stop.

1.12.4 GUILD SUMMER TOURS

Guild summer tours were held during the long vacation, and we traveled from tower to tower, everybody ringing a short "touch". One summer we rang in Oxford, with accommodation at the women's college, St. Anne's. The women were away for the summer I am afraid. This was true of a physics teaching conference I attended where we were put up at the women's dorm at MIT. It has a beautiful view over the Charles River Basin, but again, the women were absent. At Bristol, we stayed at Wills Hall. I wonder if they have no smoking there now? Wills Woodbines were favorite cigarettes of people in my childhood though I never smoked myself.

1.12.5 BALLROOM DANCING

That last year at school my mother had me take lessons in ballroom dancing, which has proved a good thing in later years. I was vastly shy, and the idea of taking a woman in my arms and dancing with her filled me with uncertainty. I went with a friend, Robin Mills, and one important worry was, what to do if we got an erection while dancing? There was some suggestion we should wear a very tight pair of underpants, so it would not be obvious.

My school suddenly decided to have joint dances with the girls' division, and since I was one of two or three who had had lessons, I was an instructor. I never did manage to date any of the girls – too shy. I had no sisters, nor did any of my friends, and the school being all male, sex was a closed book to me – until much, much later. The only time we even saw the girls at school was at lunch. The two school divisions – (boys' and girls'), were completely separate, but about a hundred yards apart. Our school, being in the process of construction, had no dining facilities, so we had to march through the girls' school in a body to the dining room. Those boys having girl-friends generally managed to see them and slip them a note at this point.

Cambridge colleges were sex segregated then too. Not so now – after 500 years, my college decided to admit women – and about time! One thing I learned about dancing is – the most important point is to make the lady look good – you don't have to be good – just look good. Hence, if you are dancing with a partner whose feet tend to get tangled in yours, you hold her so she is on your right hand side, arms high. This looks as though you know what you are doing. Then, her feet will be fully to the right of you, and you won't overlap. Further, if you do it properly, it looks fantastic. I found this is particularly important in the tango, a dance for which I never could get the steps quite right (nor did most of the women I knew!) The Hambo was another problem. This Norwegian folk dance is a mixture of a waltz and something else. My wife and I learned from different teachers, and have never been able to get it straight. We inevitably trip over one another's feet.

1.13 GETTING IN TO UNIVERSITY

My last two years at school were quite interesting. The first year was spent trying to win a scholarship to university. This was difficult, because of the competition with the demobilized servicemen, who naturally got first choice. My first attempt was at Christmas time, for Queens' College Cambridge (the Cambridge colleges had rather weird times for examinations, because they had to be taken at the college itself, so they could only occur when the students were on vacation).

Why Queens'? My chemistry master (Mr. Jarey) had been there, and I quite liked him. I took the train from Bolton to Cambridge, and had to travel sitting on my luggage most of the way. Transportation was difficult just after the war. I then took a bus to Downing Street from the railway station, walking the rest of the way – quite a walk I may say. I was greeted at Queens' by the porter in his lodge, who addressed me as "sir".

Now, I had never been addressed as "sir" in my life before by anybody at all much less a prestigious individual wearing a bowler hat (a "Derby" – at Trinity college they wore top hats!). This gave me a totally unwarranted sense of importance, from which I have yet to recover. He directed me to my rooms, which were in the old part of the college – in fact they had been built in 1448. They were exceptionally cold, and there was no heat. I could not sleep for the cold, and put the rug from the living room on the bed. This I removed and replaced in the morning, because I was afraid the bed-maker would object. I was very young.

The next day I noticed outside the room was a large wooden box, with coal in it. So that night I lit a small fire. This was shortly followed by the irate appearance of my next door neighbor, whose coal it was. He was rapidly mollified, and invited me in to sit with him in his somewhat warmer room, he now having his coal back which I had purloined. It turned out he was a spy. At the end of the war, there were many spies demobilized, who had to find some other profession, there being a small demand for spies (except for Russia and Germany perhaps). Anyway, he regaled me with tales of daring landings on the French coast from submarines using a rubber dinghy, and life abroad as a spy. It turned out there were so many ex-spies in Cambridge, in fact, that they had formed a club called the "Cloak and Dagger Society" — I wonder if it still exists?

The Sunday morning I went down to the chapel for the service. I was the only one there except for the Dean (Henry Hart) who was conducting the service and had a terrible cold. I couldn't very well back out of the communion, and he couldn't not give the service. It was a rather depressing performance, I am afraid.

The written examinations were quite terrifying. They were held in the Senate House, a very imposing eighteenth century classical building, in which was a hall with hundreds of desks at each of which sat a student. As I looked around I realized that of all these students, perhaps a dozen or so would receive scholarships. A depressing prospect to say the least.

The worst experiences were the oral examinations. I recall one in Christ's college. The room had a large plaque on one wall to say this had been Charles Darwin's lodging when he was at Cambridge. For the most part, the questions were somewhat inane. Typical was — "have you read any good books lately?" Anyway, the net result of my visit to Cambridge was — I did not get a scholarship. So my school year progressed, and finally I took the Higher School Certificate, on the basis of which I got a scholarship to Manchester University. Should I take it? I visited there, and met with Serge Tolansky, a famous physicist, who much impressed me. I had to wait in the lab next to his office, and after I had been there about ten minutes, a piece of apparatus in the middle of the room suddenly exploded with a tremendous bang. It turned out this was a cloud chamber, and when a particle of the right kind went through it, a piston inside it rapidly expanded.

The ultimate decision was — no. The reason may seem peculiar, but even now, I think it was the right one. Firstly, being so young, I would have another shot at Oxford or Cambridge. If I had gone to Manchester, I would have been unable to live there — the returned servicemen occupied most of the residential quarters. This meant I would have had to travel at least an hour and a half each way by bus — first from my home to the Manchester bus, then in Manchester, changing to the bus out to the university. However, there was no guarantee that I would get in to Cambridge! I decided to wait. My last year at school, I was able to relax — I did not have too many classes, and I was made a monitor.

Order in the school was provided by the boys themselves. There was a head boy, and a bevy of monitors, whose job it was to keep the other boys in order. To give an example, it was forbidden for boys to smoke — at least in school. The pig club (described below) had a large building, which, in the days when the cotton barons owned the place, had been a billiard room. We used it now for cooking the food for the pigs in an enormous kettle or pot, heated by gas. The cooking food made a horrendous smell, and this is why boys in need of an illicit fag (cigarette) would come here — the place was quite isolated.

However, periodically, the monitors would raid us, carting off the tobacco fiends for punishment. This punishment, carried out by the head boy, would consist of a lecture on the ills of smoking followed by stretching the culprit on a table, and swiping his rear end a number of times with a gym slipper (sneaker), while all the monitors looked on. This medieval punishment worked quite well on the whole, satisfying the sadistic instincts of the monitors, without damaging the culprit too much, and when I was promoted to monitor, I was made a confederate to this. Of course, by then the pig club was pretty well defunct, the war being over, so I was not required to turn in my friends.

1.14 THE PIG CLUB

Now what, you may ask, is a pig club? I entered form (class) Shell A1 in 1939 underage and underweight. In common with most English public schools, we were obliged to play soccer and cricket. Cricket I did not mind – as with baseball, half the time you were not doing anything, so I could read, or talk to my friends. Football was another matter. The weather in winter in Bolton was always bad, so the game was played in a sea of ice cold mud, which I found loathsome, also because I was younger and lighter than most boys in my class, I lacked the necessary coordination to play ball games. This was followed by a cold (ice cold) shower. Anybody who could survive this deserved to govern the world (as the British Empire was doing at the time – except of course for America). We then all got into a hot foot bath. This was similar to a hot tub, without the jets. It must have been a haven for bacteria, twenty or so kids in one bath. I wonder if they still do it?

Anyway, some bright spark in the class ahead of us had come to the idea that we could kill two birds with one stone. He suggested the school keep pigs. This was a very patriotic thing to do – feeding the starving (or at least ill-fed) population of England by employing left-overs from the school canteen – all that uneaten spotted dick, etc. to feed pigs. Spotted Dick was a school favorite, composed of suet pudding with raisins in it. It lay on your tummy like lead all afternoon. Pig keeping allowed us to escape games, because we would be feeding the pigs, cleaning out their sties (ugh) etc. instead. This is (or rather was!) a patriotic way of avoiding playing football and cricket, which were otherwise compulsory (playing fields of Eton, and all that). To support the war effort, we would keep pigs in the pig sties left over from an earlier generation, when the owners of the mansion (Beech House I believe), part of the school, had kept swine. Why the sties should have survived, I will never know or even why the owners of a mansion (built on the proceeds of cotton) should need pigs. Pigs are very interesting animals. Trying to catch one when it is put out of the sty for cleaning purposes is quite a problem. They are incredibly powerful, and have a tendency to run between your legs.

My colleague Critchley and I inherited this arrangement when the boys founding it graduated. The owners of Beech House had, adjacent to the building, a large billiard room (mentioned previously), with a skylight but no windows. A huge cauldron, heated by gas, had been installed in this defunct billiard room, which was at some considerable distance from the main school, and we would cook up the mess for the pigs, anything unwanted being stuck through a hole in the floor. Anyway, this worked fine during the term, when we were easily able to feed the pigs on scraps left over from the school lunch table.

Problems arose however at the end of the term, and we could not let the pigs starve, (after all, the whole idea was that they should feed us). How were we to feed these voracious pigs? So we went around the neighborhood, where special bins (called pig bins) had been set up for people to put scraps of food in, to help the war effort. The town collected these, and presumably used them for their own nefarious purposes. After we had been doing this awhile, we noticed we were being followed by a gentleman wearing a bowler hat, (a Derby – sign of bureaucracy) who would appear around corners, or out of entryways suddenly, making notes in a little black book. Soon enough, we found out his purpose. We got a summons from the headmaster ("Joe Boss" we called him – 1 have no idea why – his name was Richard) who was accompanied by the bowler hatted gentleman. We had a lecture on how wrong it was to rob the pig bins. We found food for the pigs somehow that summer – but not from the the pig bins. In the fall, we got a side of pork to divide up between us – it was very good. I have often wondered what the headmaster thought. I cannot help believing that despite his serious demeanor, he was vastly amused. It might be interesting to follow the careers of the members of the pig club. Many became eminent lawyers and doctors. Those showing a distaste for organized sport formed a small minority, since it was an attitude considered a sin at the time – as it is here in America, where many believe my institution, the University of South Carolina, was created solely to provide a football team – the American game of course – the game we played was soccer.

One other event in the war: I recall looking across Chorley New Road, which ran past the school, to see the tail of a spitfire fighter airplane sticking out of the roof of a house opposite. I have always wondered, was this the result of a former school student showing off? I suppose I shall never know. New pilots thought themselves invincible.

I had never been very proficient expressing myself in the English language – as anyone reading this can tell – however, it was necessary to communicate pretty well to get a Cambridge scholarship, so I was given special tutoring. This consisted primarily in writing essays. I find this rarer in America than in

England – but I do believe that writing is the best way to learn writing. At any rate, I had been writing an essay every week for quite a while, when I discovered that the master I was working with took my efforts (unbeknownst to me) over to his advanced English class to use as a bad example. I fumed over this for a while, then came to the conclusion this might not be too bad, because I got first class criticism free – which is, of course, what I needed.

At that time, it was necessary to learn Latin to enter Cambridge – you had to take a test. Actually, you could take Greek instead, but however difficult Latin was, Greek looked worse. So I took a crash course in Latin, which is why I have never liked the subject. I was given special tutoring. I had the option of Virgil's Eclogues and Georgics or Caesar's Gallic Wars. I picked the former. The only use I found for this ability was translating tombstones while I was resting between bouts of change ringing on church tower bells. I also took up brass rubbing for the same reason. In this occupation you place a large sheet of paper over the engraved brass plate and rub a soft pencil or a piece of cobblers wax called a "heelball." It was considered a weird thing to do then, but has since become quite popular. Now you actually have to pay to be allowed to rub a brass, most of which lie on the floor of the church, and you can even rub fake brasses in the crypt of St. Martin in the Fields Church in London.

That Christmas, I again made the trip to Cambridge, being put up in newer quarters – the part of the college built in the 1930s. I was awarded a major scholarship, of which the college gave only four. This was quite an honor, but the principal advantage was that I could now go up to Cambridge in the fall, and live in the college, in what proved to be large and imposing rooms – the Dockett building, erected in the 20s and named after Andrew Dockett, the first president of the college. Queens' is the only college to have a president. All the others having masters. Dockett would have made a good president of an American college. After the king of our first foundress (Queen Margaret of Anjou) got bumped off in the wars of the roses, Dockett convinced the new queen (Elizabeth Woodville of the opposing faction) to re-found the college.

My rooms had a coal fire, and I was obliged to carry the coal in a large sack from the dump about half a mile away. I had to put the coal sack on my back, and carry it up three flights of stairs. So much for the romantic ideas about luxury life in Cambridge.

1.15 SUMMER IN LANCASHIRE

If you have ever heard "The Lark ascending" by Vaughn Williams, or Delius' "On hearing the first cuckoo in spring", you will have some idea what summer was like walking over the moors (similar to those in Wuthering Heights) near my home in Bolton. Up towards Winter Hill, or Scotsman's

Stump (well-known landmarks), the silence was deafening – then you would hear the lark, singing its heart out as it rose higher, and higher, and higher. My friends and I would hike over Rivington Pike (a watch tower on which, ostensibly, they would light a fire on the approach of the Spanish Armada) and around the reservoirs, installed to supply Liverpool with water I believe.

One time we were cycling around and stopped at the "Yew Tree Inn". It was during the war, and the proprietor had been clearing out his cellar, coming across several bottles of genuine ginger beer, that had been laid down at the time of WWI. He allowed us to try these out, so we opened a bottle, and the liquid shot out like a cannon. We opened the next one most carefully. It tasted delicious. Old wine has nothing on old ginger beer.

1.16 DR. WRIGHT

During the whole of my life in England, I had the same doctor – Dr. Wright. He brought me into the world, and restarted my mother's heart after it stopped for nearly 3 minutes while she was having all her teeth extracted (at home in bed). He even visited me when I was sick and in bed on the way from Australia to the USA. His wife, oddly enough, had been Lord Rutherford's student in Manchester – Rutherford (the famous Nobel prize winner) taught elementary classes, as most of us have done. Once I had a polyp on my neck, which Dr. Wright cut off under local anesthetic. This he did in his "surgery" (in America – Doctors's Office) which was part of the house in which he lived. He had a small waiting room outside the surgery. Immediately after cutting the polyp off, he had to answer the phone (no nurse in those days). Unfortunately, I could feel myself fainting – the shock to my system I suppose. I bent forward to try to recover, but passed out, and slipped forward striking my head on the fender around the fireplace. It must have made a tremendous bang, and Wright came rushing in. I quickly recovered, and no harm had been done, but Wright let me out of a side door, never normally used. Then he let the next patient in, and out through the waiting room as usual. I learned later, apparently there was much discussion amongst the other patients as to what had he done with me. I had just vanished. Perhaps it was a Sweeney Todd ("The demon barber of Fleet Street") arrangement.

2 CAMBRIDGE AND QUEENS'

2.01 ARRIVAL

I went up to Cambridge in the fall of 1947, being 18 at the time. This made me very young to enter, since the bulk of the student body was composed of returned servicemen, many being married with children. Studying was very difficult for them, having been through hell during the war. I also had the advantage of rooms in college, a privilege of being a major scholar.

There was only one requirement to my scholarship – I had to say grace in Latin at dinner many nights a week. I can do this to this day at the drop of a hat.

Benedic, Domine, nos est dona tua, quae de largitate tua sumus sumpturi; et concede ut, illis salubriter nutriti, tibi debitum obsequium praestare valeamus; per Christum Dominum nostrum.

One time, in Charleston SC, at a party for Cambridge graduates, I met a lawyer who said he was also a Queens' scholar. How could we tell if we were both truthful? We recited the grace together perfectly!

One had to be careful about saying grace. The students wanted it said as rapidly as possible, so they could get on with their dinner. However, if you said it fast enough to satisfy the students, you would be dragged over the coals by the dean, (Henry Hart – a religious position at an English university) who would tell you to say it slowly, and meaningfully.

2.02 DOCKETT BUILDING

I was assigned rooms in a building erected in the nineteen twenties – designed to look like fake late Tudor architecture – what Osbert Lancaster referred to as "Stockbroker's Tudor", except it was stone and brick, not wood. It had all the usual inconveniences of English architecture – no facilities – the shower was in the basement as were the toilets and you washed in a big basin into which you poured cold water from a ewer (a big pitcher) and hot water from the kettle (there was a gas ring by the fireplace). The chamber pot (jerry) was in a cupboard under the wash hand stand, which had a fake marble top on which stood the bowl and ewer. The jerry was subject to pranks. The favorite was to place a tiny crystal of permanganate in the pot. Anyone peeing would produce purple pee – enough to trigger a heart attack.

2.03 CHAMBER POTS

The most famous chamber pot prank was performed by the Cambridge night climbers, who scaled the college buildings by night – a dangerous proceeding. A favorite climb was the spires of Kings College. One time, they left a chamber pot on the top of one of the spires. The college authorities were upset by this, but they managed to get a crack shot to shatter it. However, the next night the same climbers replaced the pot with a metal one. After much trouble, this too was shot down. The third night, the pot was replaced – this time accompanied by the union jack tied to the spire. This really presented a problem – you cannot shoot at the flag! Somehow it was removed, and the authorities stopped the climbers by inserting a concrete block between the two closely spaced buttresses where you had to climb up by putting your back on one buttress, and your feet on the other. What I believe climbers call a "chimney". Last time I was in Cambridge, these blocks were still there. No doubt these will be a puzzle to future historians unaware of the story.

2.04 TAKING A BATH

You had to cross two courtyards for a bath. In winter this meant putting on a heavy overcoat and trudging miles through the snow. Ugh! In the old days, students bathed in a hip bath in front of a roaring fire – no central heating of course. However, filling the hip bath was the job of your servant – but I didn't have a servant! One time, the British Labor Party held its annual meeting in Cambridge. Never again – the members – coal miners and such, found the privileged life the students were supposed to lead turned out to be a big myth, and shivering in cold rooms was not the laborites idea of luxury. Next year, they occupied a Brighton hotel! Of course, subsequently, Mrs. Thatcher, the then Prime Minister, almost got blown up in such a hotel. Anyway, in the old days you had a manservant, called a bed-maker, or bedder, to look after you, who would come in, tidy up, and put the kettle on to boil while you were still asleep, and set the fire. After lighting the fire you had to use a "blower" – a big thin metal sheet placed in front of the opening of the fireplace, to create a draught to encourage the fledgling flames. During and after world war one, there was a lack of men to do these somewhat menial tasks, and the job went to women. Here arose a problem. Since the college was exclusively male, the thought was that the young students, deprived of female companionship, and very horny, would take advantage of the bedder. Hence, one of the requirements of the bedders was that they should be extremely old and ugly, or at least very unattractive. Many of them were, but they were at the same time very pleasant and friendly, which meant a lot to a lower class student from a mill town such as myself, away from home for the first time.

2.05 THE CAM

Cambridge is situated on the river Cam and the river plays a big part in the life there, from swimming in it, skating and cycling on it, together with falling into it (from a punt). Originally all the sewage also went into it, and swimming was forbidden. You could see where the sewers from our 15th century building entered the Cam. In the 18th century, the Queens' Essex building was constructed directly on the river. The river bank being largely mud, this building slowly sank, until one day it leaned so far that the dean (my acquaintance Henry Hart) found his door was stuck, and he could not leave his rooms. He eventually escaped from his imprisonment, but it was clearly high time to shore the building up, and embed steel pilings in the river to stop further destruction. The river is quite shallow, and in the nineteenth century, canal boats were towed along it by horses who walked on a gravel path in the middle of the river – you can still feel the gravel with your pole while punting.

In winter the Cam often froze. I had no skates, so I decided to cycle on the ice, which was covered with a light snowfall. It proved a delightful experience – you could go fast and far with practically no effort at all, since the river is pretty straight, and the friction with the ice was very small. You are balancing on a narrow tire, so there is no tendency to fall if you go straight. Wide curves are also OK, but a sharp curve or sudden stop is a no no.

The Cam had a disastrous flood in 1947 while I was there. The Queens' new building had the ground floor under several feet of water, so you could float a punt in through the main gate. It also destroyed the expensive squash court floor.

Punting was a major summer occupation of students. The big thing was to get a punt of large heavy men poled by a thin and pretty female. My college had an antique wooden bridge, ostensibly built by Newton, but in reality not so. Standing in the middle of the bridge, you could grasp the punt pole of an unwary punter passing underneath as it was raised by the punter. This would leave him (or her) unable to control the punt unless there was a paddle, and it was the cause of much hilarity to bystanders.

2.06 THE TUBA PLAYER

Strange and interesting things happened in college. For instance, there was a student in the adjacent college, Kings, who played the tuba late at night. Since there was a stone wall between the colleges, topped by a barbed wire fence, he could not easily be stopped. So the Queens' students attached a hose to the fire hydrant, and directed it into his room through an open window, completely flooding the tuba, and everything else. Since the colleges were very jealous of their privileges, this caused an immense amount of trouble. Amongst all the others, I was called in to see the senior tutor, and under normal circumstances, I would have shared the joint guilt. However, a curious state of affairs had occurred. I love to eat asparagus, and the college grew a bed of the tastiest such delicacy. Asparagus beds improve with age, and ours was hundreds of years old. So occasionally, we would be allowed to eat this marvelous food, and this was one of those dinners. As a result I had had a surfeit, and later gone to bed sick, thus missing all the fun with the tuba, including punishment.

In many ways, the college then was like a barracks – or more properly, a lunatic asylum, in that one was allowed out, especially at night, only under very restricted conditions. One had to be in by midnight, and I well recall hearing the clock begin to strike this hour as I was in the middle of a hand of bridge in a neighboring college – Pembroke. I leaped up, and dashed out, reaching my college just in time for the last strokes, thus avoiding an unfortunate meeting with the senior tutor.

2.07 THE VANISHING WIRELESS SET

In order to have a wireless set (a radio) you were supposed to have a license, provided by the post office for a suitable fee. Being an impecunious student, I wanted to avoid this, so I designed and built a radio which fitted into a small attaché case. It was a very good, but simple, regenerative short wave receiver, which also got the long (Radio Luxembourg) and medium wave. The post office had a van which traveled around the neighborhood, with a device which could detect operating radios. My idea was, the suitcase radio did not look like a conventional set, and would not be detected. Luckily, this proved to be the case.

2.08 FRIARS BUILDING

My second and third year I occupied rooms in the dormer of Friars building a late nineteenth century construction, which I much enjoyed. It had a peaked roof, with a dormer window, in which I placed a desk, so I could both work, and look out. It had a gas fire ideal for toasting crumpets. Delicious. I had a long extensible metal toasting fork to hold the crumpets to the fire. Then Sunday afternoon, my friends and I would walk across the fields to Grantchester, a small village way in the country, returning ravenous for tea. Those ginger cakes from Fitzbillies were a delight. I note this shop, which stood at the corner of Trumpington and Downing Streets, is still there over fifty years later – but no ginger cakes. "Oh! Yet stands the church clock at ten to three? And is there honey still for tea?" (Rupert Brooke, the old Vicarage, Grantchester).

The furniture came as a separate item – I have a feeling one bought it from the prior occupant, then sold it to the next inhabitant on leaving. This made for a very interesting set, some of which I fully believe was Victorian, whereas other pieces had been purchased just last year.

The bed had sat in the same spot for a century. When the legs of the bed of the student in the next set of rooms went through the floor when he jumped in bed, having worn through the planks, and giving him quite a shock, they nailed squares of wood under each leg of my bed. Then there was the occasion when the student in the rooms beneath mine tried to gas himself. Luckily the room was leaky enough he did not succeed, and the bed-maker caught him.

The members of the college were of great variety. Many had come from abroad. (Aba Iban had been a member). As a result, I got a cold or other respiratory infection every year just after returning to college. The returnees brought back with them the most virulent strain of disease from throughout the world. Right opposite me was a student, named Quarty (or Quagrain, I forget which) from Kenya – (the other student was from Uganda). Then there were the progeny of the aristocracy, of whom I saw little, and the sons of the working class, of whom I saw a lot.

2.09 WOMEN AND PROCTORS

In some ways, Cambridge delighted in preserving the traditions of centuries. The first year I was there, women were not allowed to proceed to a degree. They attended the same courses, and took the same exams, but at the end they were given a title to a degree. Now, what the difference between a title and a degree is, I was never quite able to work out.

Having been made members of the university (which for me occurred at the end of my first year) the women had to fulfill the other requirements – one of which was to wear black gowns after dark. In order to enforce this regulation, proctors were caused to walk the streets, together with their "bulldogs". The proctors were generally elderly dons (faculty who dined on high table) and the bulldogs college servants, dressed in top hat and tail coat, and carrying a book of university regulations. Coming across an unwary undergraduate without gown, they would ask his name and college, and fine him six and eight pence (one third of a pound sterling). Caught under the same circumstances, a BA would be fined thirteen and fourpence – however, an MA did not need a gown. This was the principle reason I took my master's degree. All I had to do was pay five quid (pounds) two years after receiving my bachelor's, and I became a master!

If you tried to run, the bulldogs would chase after you and catch you, because they knew all the streets. The best plan was to brazen it out, which I did at least once. Seeing the proctor, you would walk by and nod at him as if you knew him. It took a brave student to do this, but an even braver proctor to accost you after that. There was a competition amongst the women, as soon as they became university members, to see who would be first to be "progged" as it was called. The proctors, being generally elderly bachelors, were not at all anxious to do this, and if they saw a likely prospect, would walk rapidly in the other direction. Sooner or later one woman was "progged", and after that interest waned.

I was both shy and ignorant, as far as women were concerned, but the rules of the college led to curious results. A woman in your rooms after midnight was totally forbidden. However, each set of rooms was sacrosanct, so if you "sported your oak", i.e., shut the big outer door to your rooms, no one could come in. So the woman had to stay there all night, and escape in the morning. Unfortunately this never happened to me.

2.10 KING'S CAROL SERVICE

King's college had this famous choir which performed the Christmas carol service – Nine Lessons and Carols. After the war, the carol service resumed, but the chapel was not heated. There was no problem about getting tickets! I would put on my pajamas (maybe two pairs) my thickest shirt and pullovers, coat, overcoat and muffler.

Then I would fill a hot water bottle, and walk over to King's chapel, nearby. Feet on the hot water bottle made it bearable – but the service was just as ethereal as it is today. I generally went to the Advent carol service, as did most other students. We were home at Christmas.

2.10.1 THE QUINCENTENERY

In 1948, the college had its five-hundredth birthday. The queen mother, being the patron, came down for the celebration, and there was quite a fuss. I recall my supervisor in chemistry, a Dr. Ramsey, was in charge of the dinner, and had to discover which wine the queen liked. It was a white wine. A garden party was held, and the queen came and walked and talked to people. After her death at an advanced age, the present queen took on the chore of being patron. I am not quite sure what a patron does, other than looking after the interests of the college.

2.11 KING'S CHORISTERS

The treble choristers at King's were young boys, who attended the choral school nearby. They dressed in top hats and Eton jackets, together with an Eton collar (A large, wide uncomfortable white celluloid collar). One time I was leaving my college by the back entrance, wearing my gown (as mentioned earlier, one had to wear a gown after dark or get "progged"). The choristers in a "crocodile", two by two happened to be walking past, blocking my exit, and each pair raised their top hats as they passed me. A rare experience.

2.11 COLLEGE LIFE

College life at Cambridge was different from a normal existence. In many respects it had not changed in 500 years – the totally male environment for example – it came as quite a surprise on returning in the eighties to see girls, now college members, sitting reading by the Cam in the Fellows' Garden – a good thing too I believe! Had there been women around when I was in residence, I might well have stayed!

2.12 DINNER IN COLLEGE

Dinner in college was interesting. The bare tables had long benches to sit at. You had to climb over the table to sit against the wall. The tables filled up from the head down – Hobson's choice. Hence, you never knew whom you were sitting with (or with whom you were sitting!).

The rowers were the worst. They could talk about nothing but their sliding seats. I found those who were law students, and who intended being involved at court (with the royal family) most intriguing, because their lifestyle was so different from mine. I learned that although Princess Elizabeth was fine, Princess Margaret Rose was a real so-and-so. This was, of course, complete hearsay, but I think there was a certain amount of truth in it. Later events proved this judgment probably correct! Still – de mortuis nil nisi bonum.

2.13 THE FOOT-IN-THE TOILET PRANK

The fact we all sat adjacent to people we might well never have met before led to interesting situations. On one occasion, the students I sat next to (and myself) came up with a bright idea. One of us had broken his leg at the ankle – I forget how, tripped and fell maybe. We had been arguing that if you all agreed with one another, and tried hard enough, you could convince people of complete impossibilities. Anyway, we all agreed to convince the next person to sit down with us that the leg had been broken by jamming it down the toilet bowl. Of course, we had to come up with a splendid plausible reason why this might have happened – perhaps he was pushing something down the bowl with his foot, or reaching for soap on the toilet shelf. Anyway, if you come up with enough plausible false detail it is amazing how easily you can convince people. To this day, I am sure the next two people who sat with us are convinced it happened the way we described. I myself am subject to this. I recall Arnold Jones, my colleague, convincing me that our big resolving magnet on the accelerator was cracked – and I believed him!

2.14 MEDICAL STUDENTS

We had many medical students in college, and sitting with them was a hazard. We still had food rationing at this time, and the med students would describe the most gory, grisly operations with the idea we would lose our appetites and they would have more to eat. They also had the most bizarre ideas too. I recall one student had sewn buttons around the waist of a cadaver, and convinced a very naive nurse that this was the way men held up their pants.

2.15 STUDENT PRANKS

One thing about Cambridge was the extreme inventiveness of such student pranks. For example, a group of students dressed as navvies (manual laborers), dug up the road, placing a barrier and large signs saying "men at work" etc. They then left. It was several weeks before people realized this was not a real repair. On another occasion, one Indian student dressed as a Maharajah, and attended by a retinue of suitably attired servants, convinced one of the smaller colleges to invite them all to dinner. Then there was the time students put it about they were going to burn down the stalls which were erected in the market square. The stalls consisted of a canvas roof on wooden poles held in metal sockets set in the paved market place. The town, thinking ahead, took down the stalls. The students, under cover of darkness, then filled the sockets with concrete.

2.16 EVEREST

Mount Everest was first climbed in 1953. I would have dinner in college with a student who was on the expedition. He got as far as the second or third camp from the top. I recall he said the most important piece of equipment he took with him was his umbrella. Apparently at that time of year, there is much rain on Everest.

2.17. CHEMISTRY, MATHEMATICS AND CRYSTALLOGRAPHY

Although majoring in Physics, I had to take a number of subsidiary subjects. The principal one was chemistry. I was taught organic chemistry by Dr. Kipping. This was a descendent of the original Kipping, of the "Perkin and Kipping" organic chemistry book. I recall his bringing in a lump of silicone he had just been sent from the United States. This was what we would now call "silly putty", but it was a fascinating substance then. An American post doc was the organic lab TA. After we had synthesized one substance, he pointed out this was "Spanish fly", the aphrodisiac, but we did not consume it. We did not rely on our synthesis!

Prof. Norrish lectured on physical chemistry, which was a requirement of his professorship, but he was not particularly keen on this, and generally left it to a subsidiary – Dr. Sugden – (not the parachor man).

The chemistry labs on Downing street were somewhat antique. Students would pour ether down the drains, which were partly glass tubes hung from the ceiling below. One time, someone threw a lit match in a sink, which ignited the ether, and the glass tubes exploded, sequentially, one after the other, bang, bang, bang, a remarkable sight.

I had a tutor in chemistry, a Dr. Allen Sharpe I believe. I recall that he had to have every tooth in his head removed in the middle of the semester, because they had rotted away. It turned out he was a fluorine chemist, and the fluorine had got at his teeth. This seemed odd to me, because of course, fluorine is used as a protection against tooth decay – but evidently only in small amounts.

2.18 BLOWING GLASS

One of the requirements as a physics undergraduate at the Cavendish was that every student must be capable of blowing glass. The primary test consisted of making a T piece of soda glass tubing, annealing it, and then dropping it on the floor. Soda glass is more difficult to anneal than pyrex, and the test was that the T piece must not break at the joint when dropped. This showed it was strong, and properly annealed. It did not matter should the T piece break elsewhere. I enjoyed glassblowing, and used to make "dragoyles" – an ether filled glass bulb with a tube leading down from it, terminating in an "s" shape. The bulb was cooled by placing a wet rag over it and bubbles rose up the tube, having evaporated in the "s" shaped region, condensing in the bulb. In making the device, it was necessary to keep the ether boiling in the bulb as you heat-sealed the tube – a hazardous business – I am sure it would be illegal today.

2.19 EXPLOSION OF THE HELIUM LIQUEFIER

One day, my colleague and I were working in the upstairs lab. I was taking data, and he was looking out of the window at the Monde Laboratory, when there came a loud bang, and a look of extreme surprise suffused his face. "My God" he said. "I have just seen the liquefier go up through the roof of the Monde labs". Perhaps this was not too great a surprise, because the Monde was designed with a very thin roof, so that, should there be an explosion, it would in fact go through the roof rather than hurt people in the next room. In fact no one got injured.

2.20 FINAL EXAM.

My final exam came at the end of the Lent Semester in 1950. Much depended on the outcome, so I was delighted to receive a telegram from Prof. David Schoenberg to say I had been given first class honors. This meant I would be accepted as a graduate student at the Cavendish Lab, my aim in life. I went up early in the fall to arrange accommodation, which I did, taking a room at a house in Owlstone road, not too far from Queens'. It turned out the place was run by a termagant, who kicked me out a month later for having put my stocking feet up on the mantelpiece in my room, to keep warm. I took a room in the basement of a house in Malcomb street which I found very congenial, a three story early nineteenth century brick row house in the heart of Cambridge. The lady running the place was a motherly type, her husband had had to quit his job as a technician for medical reasons. The occupants of the house were varied and interesting. For instance there was the Tokyo Librarian. I had long discussions with him about the different thinking process between Japanese and English. The Japanese do not employ sentences as we in the west do. Then there was the Siamese princess, who showed me pictures of herself and her pet cheetah (or was it a puma?) on a leash She seemed to have a plethora of the most amazing and unusual men friends. I spent the whole of my remaining time at Cambridge in these cozy digs, a happy life.

3. THE CAVENDISH

3.01 THE LAB

The Cavendish Laboratory, named after the seventh Duke of Devonshire, was founded in 1871. Examining a picture of the duke, he resembles a prime example of mid-Victorian rectitude (and stuffiness) and of course, he was! Nevertheless, he was also very bright. He was a wrangler (top mathematician) in his class, and went on to run the Cavendish family estates. He would not let his son follow in his mathematical footsteps, because he said a degree in math was no way to prepare for running a business.

The Cavendish was probably <u>the</u> place to be for experimental physics in the 30s and to a large extent still is today, although the competition from other universities, such as Cal Tech and MIT, is now much fiercer. Much has been written about the early period. After all, the electron, the neutron and the nucleus were all discovered there. My boss, Otto Frisch, had coined the name "fission" when he and his aunt, Lisa Meitner, wrote the first paper on it. There were, however, many curious anachronisms which still persisted while I was there. One was that one knocked off work at 6 PM (or earlier) and went home. This presented a problem if you were in the middle of an experiment, so you borrowed a key to the front door from another graduate student, and copied it. Since this was the eighth or ninth sequential copy, the key did not work too well. You applied it to the immense wooden front door, suitable for a castle or a cathedral, in fact there was the Devonshire family motto engraved in stone over the door, "Cavendo Tutus", meaning "safe by being cautious", (very appropriate, no doubt) and fiddled and fiddled. Suddenly a red face would appear out of a window directly above the door, and a hoarse voice would say, "Ere, wot are you doing down there?". This was the porter, who was in charge of the building. He always poked his head out with a bowler hat (Derby) on, and rumor had it that he slept that way. Anyway, one shamefacedly said that there was an experiment running that needed tending, and he would vanish. He played a part in the great shearwater caper, about which more later.

3.02 THE MAXWELL LECTURE THEATRE

All the elementary lectures occurred in the Maxwell lecture room. This had been designed by James Clerk Maxwell in the early 1870s, and was ideal for its period. The seats were long benches of wood, hard and straight, so that they could accommodate the maximum number of students, and were uncomfortable enough to keep you awake. The back of the benches formed a ledge for the notebooks of the people behind, and the auditorium was strongly raked, so the students at the back could look down on the lecture bench, and see what was going on. Behind the lecturer were several large blackboards, and two doors which led to a sizeable preparation room behind them. Each lecturer had a technician to prepare his demonstrations, which were often very good. The windows had wooden shutters which could be cranked closed to prevent the entry of any light. The original gas fishtail burners were still present, by which the lecture room had been illuminated before electricity, and even before gas mantles, and the original supports for the first electric lights were present. As I recall, the two wires from each carbon filament bulb were hooked over what looked like cup hooks – no plugs. Maxwell had carefully arranged that latecomers could enter from the back without disturbing the rest of the audience

3.03 THE WHALE

There were many odd things about experiments in the Cavendish. The work on hemoglobin and similar molecules required some weird materials. It had been found that deep diving mammals, seals, whales, etc., were ideal sources for myoglobin. One time a whale had died somewhere on the east coast, and we had been told we could have it if we could get it. Lab members of the Cambridge gliding club came up with the idea of using a glider trailer – what you put the glider in when it comes down – to haul the whale. On the way back, they were caught speeding. The policeman said "What have you got in there?" "A dead whale!". Of course he didn't believe them, but on opening the trailer, the smell of the rapidly decomposing whale convinced him – so he gave them an escort home! What happened to the whale later I have no idea. I remember seeing a student carrying around a complete dead penguin too – frozen – I never saw it being dissected, however.

I recall considerable quantities of radioactivity had been spilt on our shelves over the years, and one of the janitorial staff had been given the job of cleaning this up. As he happily swabbed the shelves down with a wet cloth, we said "aren't you going to wear rubber gloves or something to protect against the radioactivity?" "Oh no," he said "The water kills the radioactivity". We two lowly graduate students looked at one another. In the end we did nothing – after all, it was a low level of activity, but I still worry we should have said something.

There was a paper sack in the fume hood in our office. It turned out it was filled with beryllium oxide. There was a small hole in the bag, and in cleaning up, our assistant would blow the white powder away. After this had gone on for six months, we read an article which pointed out that a microscopic amount of BeO – less than a microgram – could kill you. The bag suddenly vanished, I have a suspicion it went into the trash. Luckily, some people are not susceptible to this poison, which works like silicosis, and thankfully, I must belong to this class. The assistant who dealt with it was no youngster, and had worked in all manner of hazardous fields, with no apparent ill effects. He was always pointed out as the example of the fact that physics was not really dangerous. Unfortunately, about six months later he died of cancer, after which his name was rarely mentioned in this context!

3.04 CLEANING UP RUTHERFORD'S LAB

Rutherford had died in 1937. He employed a laboratory at the top of a tower accessed by a narrow winding staircase which had been locked up immediately after his death. The war came, and this lab remained closed until I got there. As a lowly graduate student, a Mr. Baxter and I were assigned to clean up this room together. We entered the room with some trepidation – it was quite eerie. Everything was covered with dust, and all the glassware had turned blue. Why this should be I have no idea, but we speculated it might have been the radioactivity, which our Geiger counters showed was considerable. We cleaned up the place as best we could, but I don't believe it was used for nuclear experiments again.

3.05 CLUBS AND SOCIETIES

Cambridge was full of rather unusual societies. I have already mentioned the Cloak and Dagger club for ex-spies. The two societies most interesting to physicists and astronomers were the Kapitza club, and the Del squared V club (written $\nabla^2 V$). The Kapitza club was named after the Russian physicist Peter of that name who had returned home just before WWII and been kidnapped by Stalin.

The minutes of the Del Squared V Club are among the Archives at Cambridge University Library. They form part of the Janus reference system, and are available for consultation. They are handwritten, and contain not only the title and name of the speaker, but also a brief summary of the talk. The minutes of the Kapitza Club on the other hand reside with the papers of

Cockcroft at the Archives Centre, Churchill College, Cambridge. They are rather brief, and also handwritten, giving the title, name of the speaker and not much else. In both cases the handwriting is not always very legible. Nevertheless, one finds interesting remarks therein. For example, a vote was held at one meeting as to whether Compton was correct in his new explanation of the scattering of gamma rays by electrons. The vote was about 50-50, with Kapitza voting against.

The clubs were quite small, and one had to be elected; I forget how I got into the Kapitza club. Even J R Oppenheimer had difficulty getting elected in 1926, and in 1936 Fred Hoyle was rejected twice in getting elected to the del squared V. I recall I was lacking one vote for this club. The eminent statistician Sir Ronald Fisher was sitting on the front row. He was very short-sighted, and I had been handing him my notes to read during the lectures on nuclear physics we both attended and sat through together (I had no idea who he was then). Oddly enough, my first wife knew his family in London well, where he had a reputation for absent mindedness. He had a large number of children, and the tale went that his wife left him to put them to bed one night, returning to find he had also put one of the neighbor children to bed as well. Anyway, as I was one short to get elected to this society, he peered at me closely, realized he knew me vaguely and voted me in. These small societies were where the latest in physics was discussed. I remember well Fred Hoyle returning from a summer spent at the California Institute of Technology to tell us that "During the last summer, the universe has doubled in size". What had happened was that they had discovered that there were two kinds of Cepheid variable star, not one as had been thought previously. The star had been used as a "standard candle" – a star which was supposed to give out the same amount of light wherever it was in the universe. When it was found there were two kinds of such stars, it had led to a recalibration of distances in the universe, with the result that it now was twice as large as the previous estimation had indicated.

John Jelley gave a talk on cosmic rays. Most cosmic ray experimentalists had their equipment on the top of snow clad mountains, as high as possible. However, Jelley was examining the Cerenkov radiation in the atmosphere. Each energetic cosmic ray particle produced a cone of light radiation in the air. Jelley detected this with a large parabolic mirror having a photomultiplier – a light detector, at its focus. Hence he pointed out that the best place for his experiments was the middle of the Sahara! In America I obtained a searchlight mirror I placed on the roof of our seventh floor building to see such radiation. However, it was stolen, I still wonder how.

3.06 LECTURES

Lectures at the Cavendish just after the war were interesting. One put on all the clothes one could – overcoat, scarves etc., and took down lecture notes wearing gloves. The Cavendish lecture halls had no heat at all, and otherwise one froze. It is difficult to imagine now, in this day and age, presently famous and eminent physicists taking notes in woolly gloves – it is not easy to write in wooly gloves! Radcliffe gave exceptionally clear lectures. He locked the door precisely at 9:00 am – even the backdoor (which, as mentioned earlier, Maxwell had thoughtfully arranged for latecomers to sneak in). In spite of the clarity of his lectures on electricity and magnetism, I never could remember what he said. It seemed to me that the lecturers that bumbled along, making mistakes and correcting them, may have had certain advantages – you had to keep up with them to understand the corrections.

Lecture demonstrations were a significant part of the Cavendish style. They had a magnificent collection of equipment. Each lecturer had a demonstration assistant, and they set everything up. Mr. Crowe was the doyen. He had worked for Rutherford in the early days, and had demonstrated x-rays by putting his hand in the beam quite often. As a result, the irradiated hand looked like marble – rather strange to observe so it might have done well in a horror movie

Bragg was a very good lecturer. He had these yellowed notes, from which he taught us optics. But he had never mastered vector notation, so he would write out vector differential equations in full, a rather tedious experience. Nevertheless, he possessed physical intuition to a remarkable degree – and if you ask what is physical intuition, it is like God – you can't really explain it – but you can tell it when it is there.

Mr. E.S. Shire had been at the Cavendish for a very long time. So long, in fact, that the Ph.D. degree arrived after he would normally have taken it – so he took pride in the fact that he had not got one then. He had just constructed the Van de Graaff accelerator in a tower in the Cavendish courtyard when I started research, so at a later date it was suggested that we compare the proton and neutron yields from ^{10}B bombarded by alpha particles from the accelerator, to see whether the nuclear neutron and proton forces were the same, (apart from the Coulomb effect of course). I had developed the neutron counter necessary for this experiment. It employed the miserable Szilard-Chalmers reaction in potassium permanganate, which stained everything it touched. The experiment went very well. Shire wrote it up with my help, and I read the manuscript to check all was OK. About five years later, I was in Australia at the Australian National University, when a colleague (Al Mann I think) pointed out that the ratio of protons to neutrons was given twice in the paper, and one was the reciprocal of the other! I think we worked it out, but it proves one thing. I do not believe I have ever written a paper that is completely correct. It may only be a word misspelled, but there is always something!

3.07 STARTING AT THE CAVENDISH

As a graduate student, my first work was to design and construct a flat pancake-shaped Geiger counter to detect decay electrons from Mn^{56} deposited on a filter paper. I had to blow the glass terminals with tungsten wire in them, attach the fine wire down the middle, turn the brass casing and fasten it all together with black wax. Then I evacuated it and filled it with argon, alcohol and CO_2. You really knew how things worked when you had finished! Cockcroft and Walton designed the first high tension set used for nuclear disintegration at the Cavendish in the early thirties (parts of their equipment – the big glass accelerator tubes – were still around in the room over the cyclotron when I was there). The tradition of string and sealing wax persisted.

3.08 HT2

After Cockcroft and Walton's machine, Philips' Gloelampenfabriken in Holland started to produce commercial high voltage stacks employing enormous vacuum tube rectifiers. These look like the sort of equipment in the Frankenstein movies of the same period. The two stacks in the Austin wing hall were of this kind – the selenium rectifiers came later. The smaller stack produced about 1 million volts, and ostensibly the larger 2MV – except it never did. Big cloth "sails" around the terminal were supposed to distribute the voltage gradient more evenly – but they worked minimally, and we rarely had them installed. When they were, they billowed out like a ship in full sail, forced by the draught of air from the high tension terminal.

Rumor had it that an incendiary bomb had fallen on the hall in the war, but it fell in a sink, and with great presence of mind, someone turned on the tap. Of course, if the bomb had exploded this was the worst thing you could do, so it was probably a dud.

The particles from the accelerator in the large hall came down through a tube in the roof of the control room below, and were deflected through 90° (to travel horizontally) by an iron cored electromagnet, between the poles of which sat the vacuum magnet box. It was exceedingly difficult to remove this, which was a continuing problem because of the perennial leak which it had. The standard practice was to paint glyptal, a mysterious bright red paint, over the leaks to stop them up, but if they were where the magnet butted against the box, this was impossible, so then we would spray (and ultimately, pour) acetone down the side of the box, dissolving the glyptal previously applied, and hopefully stopping the leak. Ultimately, we removed the box, and took off the paint. We found that, before WWII, the leak-stopping paint came in various colors, so students could see which leak they had attacked – as we removed each layer, the box resembled Joseph's coat of many colors. Charge built up on various parts of the electrode system of the big stack, causing breakdown, which sounded and looked (and really was) like a lightning flash.

3.09 THE SPARK – ST. ELMOS FIRE

In order to discover where the discharge (rather like the St. Elmo's fire seen on old sailing ship spars in a thunderstorm) came from, being the lowliest graduate student, I had to go up in the gallery surrounding this machine, protected only by a very thin metal railing, The other experimentalists would turn down the lights, and I would watch for the flash as they increased the voltage on the stack. The only problem was that I jumped so much at the bang, I couldn't see where the flash came from. Luckily, I had a box brownie camera I had obtained by sending in corn flakes package tops. I set this up and opened the shutter, (it had a time setting) then developed the film in a tea cup, and we got very good pictures of the flash.

On another occasion, my coworker, John Carver, fell through a hole in the floor. Actually, he fell in and supported himself with his hands on the edge of the hole. There were two holes in our floor, which sat on a huge tank of transformer oil, acquired by Mark Oliphant with the idea of building an oil insulated high voltage machine. One hole was like 30 ft. deep, and the other about two feet deep. Carver gingerly extracted himself from the hole – it was the two feet deep one! HT1 had a porcelain accelerator tube, from which a chunk had been knocked out by a big spark at some earlier time. This was stuck back on with the universal material for vacuum repairs, Q compound. The General Electric Company had made a number of curious oils, greases, waxes and solids which had low vapor pressures. They were obtained by distilling ordinary oils, and the fraction not distilling, of course had low vapor pressure. The oils were L and M, and the solid, which incorporated graphite I believe, was Q. It much resembled modeling clay, or plasticine. Anyway, you would be running on HT1, increasing the voltage, when suddenly there would be an immense bang, produced by a spark jumping between the wall and the Q compound, and the mechanical pumps would start going kechunk kechunk kechunk, indicating they were pumping air – there was a hole in the system. Immediately you would dash upstairs locate the chunk of porcelain blown off by the discharge and replace it, smoothing the Q compound around the patch. The pumps would pick up again and go "hard" and one would nonchalantly continue the run. Actually, it was great fun, and it is difficult to describe the enjoyment one got as the results accumulated. One person did everything, set up counters, ran the accelerator, and took data. Of course, our technicians kept the machines going – but they must have been poorly paid. I recall one arguing whether he could afford an autocycle – he cycled in at least ten miles every day, in all weathers.

The large rectifying vacuum tubes burned out periodically, and had to be replaced. This involved climbing a very tall rickety ladder, and reaching out as far as one could to unscrew the extremely heavy tube, replacing it with another very heavy tube. If the tubes were not operating properly, charge could accumulate on one electrode. I well recall our technician saying as he pointed "That's where it is". Immediately, as if in response, a bolt of charge came out and struck him on the finger, semi-paralyzing it with the result he walked around pointing with his finger stuck out for a week until the effect wore off.

At one period we only ran at night, I forget why. Anyway, it turned out that unbeknownst to us, the RF used to excite our ion source was tuned to the same frequency as the BBC (British Broadcasting Corporation) television thus interfering with viewers for miles about. The BBC had a van which went around locating such interference, but as mentioned earlier, you were not allowed on the site after dark. Then one time we ran in the day, and this irate person showed up to say we were ruining peoples' pictures. Of course it was an easy thing to do to shift the frequency a little and cure the problem.

3.10 NUJOL

When phosphors and photomultipliers first came in, the information from America said you attached the crystal to the photomultiplier with a little Nujol. We, in England, had no idea what this exotic substance was, and it took quite a while to discover it was merely what the British call paraffin oil (kerosene). Some of the crystals, particularly the organic ones such as naphthalene were very fragile. One unfortunate student had got a new one. He held it in the palm of his hand to examine it closely and went "ah". As he said this, it cracked! The unequal expansion produced by his hot breath had done it.

Rutherford had been alive as the HT sets were built. The control room was miles from the toilet, and one could not leave the machine to answer the call of nature. However, there was a sink conveniently placed in a corner of the room. The tale went that when Rutherford was examining the control room, someone pointed out the sink, ostensibly there for chemical purposes. He went up, stood against it and said "Yes that's about the right height". Needless to say, at that time we had no women research students in the nuclear group!

We had the very first multichannel pulse height analyzer not constructed of a plethora of single-channel analyzers. It was invented by my thesis advisor Denys Wilkinson. This machine ran ninety nine post office registers. These were mechanical counters operated by small electromagnets. It made an interesting noise with all these things clicking. It was, however, a lash-up, and the slightest nearby interference would send it wild. Cars ran through the passage adjacent to the counting room, and they would ruin our results. Since we were only interested in the results, not improving the device, we attached strings close to the low roof all over the room, connected to the switch to turn off the analyzer, rather like the cord running down the middle of the old trolley cars which rang a bell near the driver. As we heard a car in the distance, (which happened not too often), we pulled the string to turn off the equipment. It worked fine!

We were only given a few days for our experiments, so we ran day and night. Unlike today, there were only one or two people on each experiment. I recall we had been running three days without sleep, and at five in the morning I went to the counting room, which unlike the control room was very quiet. As I took down the data I went to sleep – my pen ran off the paper. I returned to the control room, and my colleague, John Carver, got real mad "You nitwit" he said, "why the heck didn't you stay awake, I'll take the data next time", and so he did, and – you guessed it – he fell asleep too.

The hall of the accelerator was very large – and one student (Raphael Littauer, later of Cornell) produced a golf ball and took a shot with a club. The ball ricocheted all over the place, but luckily no damage was done.

One of my fellow graduate students was much prone to swearing (Adrian Cohen). He would really let fly at the least opportunity. One day his delicate apparatus, over which he had spent much time and effort fell over – but he couldn't say a thing – he had used up all his swear words on minor accidents, and there was nothing left!

We had one mad American by the name of McCutchen. He would work on petrol (gasoline) powered model helicopters in the lab, making an almighty noise, with his girlfriend sitting placidly knitting nearby.

3.11 DOUGLAS COLVIN

I shared an office with Douglas Colvin, who had graduated from St. Andrews University and had a strong Scottish brogue. He developed a large pressurized ionization chamber, filled with highly purified deuterium gas. With this he studied the photodisintegration of the deuteron using gamma rays produced by the accelerator. The deuterium was purified by passing over metallic sodium, and the problem was how to get rid of the sodium later. My colleague Carver enjoyed filling a sink with water, and tossing little bits of sodium in it. This highly dangerous and dicey procedure produced fire and a nasty smell, as the sodium burned and spun around on the water, but nevertheless, it was quite spectacular.

Douglas had a society wedding, marrying the daughter of the proprietor of the Sainsbury's chain of supermarkets. I attended the wedding, driving down on my motorcycle, covering my suit with a waterproof outfit, which I removed when I got there (Sussex I believe) and hiding the bike under a hedge. The wedding party was all dolled up in top hats etc. Later on, somehow, Colvin became a hippie for a while, I was told – but by then I had completely lost touch. I recently read his obituary and it appears he was very involved with the peaceful use of atomic energy.

3.11.1 ALAN SHARPE AND OTHER TUTORS

Teaching at Cambridge involved a tutorial system. Graduate students and young professors were paid to tutor a group of two or three undergraduates. This was an excellent method to teach, because it involved direct contact with the tutor, unlike the lectures which might involve a hundred or more students with one lecturer. Unfortunately, it was extremely expensive too.

One teacher I had was a fluorine chemist by the name of Allan G. Sharpe. As I mentioned before, although fluorine is extremely beneficial to teeth in small quantities, in large quantities it destroys them, which is what happened to Sharpe, unfortunately. In the middle of the semester he had to have all his teeth pulled. This was a disaster to students such as me, though he did try to fix us up with other tutors. Eventually he returned, but teaching without teeth is quite a hassle. I do recall he had one interesting remark. He pointed out that there was not much use in doing homework because, if you know the stuff, why bother to write it down? If you don't know the stuff, you can't write it down, anyway. I have always puzzled over that.

Another tutor I recollect was Shepard, a mathematician. In addition to his accomplishments as a prime mathematician, he was an excellent juggler. He would come to my rooms, pick up an ink bottle, a glass of beer, whatever, and juggle them without missing a beat. The first time he did this, I was aghast. Many other tutors left less of an imprint.

3.12 PERMANGANATE DETECTOR

I was given, as a research project, what was without doubt the most miserable experiment imaginable. Some years previously, Leo Szilard (whose name is famous as having induced Einstein to bring to president Roosevelt's attention the possibility of an atom bomb) had discovered that the radiation produced (gamma rays) by the absorption of slow neutrons could destroy certain molecules, producing a more stable molecule. In one such reaction, a neutron was absorbed by a manganese atom in potassium permanganate. This immediately disintegrated, producing manganese dioxide as a precipitate, which could then be filtered off. The radioactive manganese having a half-life of two and a half hours, could be counted by a Geiger counter. The advantage of this was that the small amount of radioactivity could easily be concentrated by pressure filtering onto a filter paper. The disadvantage was that processing large quantities of permanganate was the most messy, miserable experiment imaginable. I developed this process to make it as automatic as possible, but the concentrated permanganate still got everywhere, leaving a dark brown stain on clothes and skin. This could be removed with hydrogen peroxide and dilute acid – sulphuric, as I recall.

Anyway, over the years I got heartily sick of this stuff – but I did use it for a variety of experiments. At first, I used to filter a relatively small quantity, a liter or two, and count it under Geiger counters of my own construction. My earliest counters were, flat, banjo shaped, but it proved better to use aluminum cylinders, turned very thin on the lathe, and wrap the filter paper around them. I made an automatic counting device. It would count for a while, then record the number of counts in binary on teledeltos paper. This mysterious substance has long been relegated to history, I believe, but it was a paper coated on one side with something like potassium nitrate, and conducting on the other. When a voltage was applied to the paper through a wire, the nitrate burnt off, leaving a black mark. Every ten minutes or so, when I recorded data, the recorder would explode. Current would discharge through the wires with a loud bang, and smoke smelling strongly of fireworks would rise from the equipment. I used to supervise undergraduates in the same room, and if I forgot to tell them about it, when all this happened, it would put the fear of God in them – because otherwise the room was very quiet. I used this to measure neutron cross sections of photodisintegration for beryllium, and later (n2n) reactions. (No need to go into that). However, it was decided to measure neutrons from the spontaneous fission of natural uranium this way. I borrowed a large chunk of uranium from Dr. Maddox, a famous radio-chemist. It turned out he went to Brazil to give a talk a couple of days later, and the press had big headlines "NUCLEAR PHYSICIST DEFECTS TO BRAZIL WITH FISSIONABLE MATERIAL". A few days later he returned, and the press published an apology in small type. I had the uranium, and he had merely gone to Brazil to give a scheduled talk.

The Giant Dipole Resonance in nuclei had just been discovered, but there was a feeling that it was not a genuine resonance, but arose because of the competition between the n and 2n production, the n having, of course a lower energy threshold. The nuclear product of these two reactions was different, so nobody could tell if there really was a resonance, in which case the single neutron reaction would rise and fall as the disintegrating gamma ray energy increased, or if it was merely the two n reaction, which robbed the single n reaction as it got above threshold. Anyway, for various reasons, we decided to study tantalum, a rare metal at that time. We got a metal box of this substance from the Johnson Mathie Company, and put the box in the gamma ray beam coming from our proton target. Then I put my neutron detector close to the box, and we measured our results without ever opening it. When we returned the box, and told Johnson Mathie we had not opened it but got good results, they were quite mystified.

3.13 CAMBRIDGE DIGITAL COMPUTER

It is interesting how time catches up with us. At Cambridge in 1950 or 51, they had erected the first digital computer – at least in England. It occupied a fairly large room on the second floor of a building utilized by the mathematics department, in which they had previously had a large number of mechanical calculating engines; primarily ball-and-disc integrators. The room had in it rack upon rack of tube-type electronics, which generated so much heat that a large exhaust fan filled one window – a very rare necessity in England's cold climate.

My colleague Arnold Jones and I were encouraged to use this device, but we were uncertain whether or not it would be an advantage. Don't forget this was before there were such things as programing languages – you had to start from scratch. We decided to have a competition – Arnold would use the computer, and I would do the same calculations using the mechanical calculator – a Friden, I believe, where the additions were performed by rotating cog wheels. I won hands down, simply because it took so long to program even the relatively simple calculations involved; running them was very rapid. The other day, I received in the mail a catalog of reprints of ancient and historic physics texts – for instance, Newton's Principia was there, and his Optics. Lo and behold, there was also the first manual on computing language ever written, by the director of the computer we worked on, who was also the guy who taught me statistics. His name was Maurice Wilkes. Now that is just so much history. Makes you feel old to be classed with Newton!

One of the most interesting features of the old machines was their memory. Unlike modern computers where storage can be on magnetic tape optical disk or integrated silicon devices, the memory used then was transient – it consisted of an acoustic "delay line" – a tube filled with mercury, pulses being fed in at one end acoustically, taken out at the other and recycled. Now, if the computer went down, because of tube (valve) failure, which occurred every one to ten thousand hours per tube, the whole calculation was lost. As a result, it was common practice never to run a calculation for more than ten minutes. The results would then be taken out of the machine, and the calculation continued. This proved quite cumbersome, but was indispensable, because if you have a computer with ten thousand tubes, each lasting 1000 hours, one tube burns out every six minutes.

The same type of memory was used in pulse height analyzers, or "kicksorters" as they were called then. I well remember Tommy Gold, who was an eminent astrophysicist, designed such a one using plastic tubes about a meter long filled with mercury. After its construction, we all went home for

Christmas. However, unbeknownst to us, the heat in the lab was turned off to conserve energy during the Christmas break. On returning to the lab in early January, we found a vast mess of mercury on the floor of the workshop, which took quite a while to clean up. What had happened was that the differential thermal expansion of the plastic tubes containing the mercury, and the metal frame supporting the tubes, which also held the "mirrors" reflecting the ultrasound in the mercury column as well as the microphone and speaker, had produced a leak. Coils of nickel wire delay lines replaced the mercury as memory, causing our mechanics to heave a sigh of relief. Considering the amount of mercury found in floor cracks of the old Cavendish, it is surprising we did not all end up with hatter's madness (poisoning caused by mercury and its vapor).

The nickel delay lines also produced problems. Hutchinson and Scarrot, who had designed the delay line process, had used nickel wire they found stored in the basement of the lab. The magnetostriction in this worked fine, and they wrote up the device in the journals. They then received letters to say no-one else could get it to work. It turned out this wire was a special batch, with nothing else like it. Eventually more wire was found which worked.

3.14 NEUTRON STANDARD CALIBRATION

At that time there was a large discrepancy between neutron standards – the European and US standard sources differed by about a factor of two! That is to say, if you measured the output from one source using British techniques and do the same using American methods, the British would say one thing, but the US twice as much. We wished to make a device based on fundamental principles to provide an accurate calibration

The experiment was interesting, but not the technique used. I had a vast tank (500 gallons) of concentrated potassium permanganate. The tank had been acquired under curious circumstances from a spiv in the East End of London – I visited a junk yard there, where they had cut the top off a second hand steel tank, and coated it inside with polyethylene plastic using a gun device. The neutron source was placed at the center of the tank, which was filled with permanganate solution, and the radioactive manganese dioxide produced by the Szilard Chalmers reaction of the slow neutrons on the permanganate filtered off as mentioned previously. It was an exceptionally messy experiment, which I got heartily sick of, but it did allow an accurate comparison of neutron sources. A source placed in the middle of the tank would have a large fraction (80–90%) of its neutrons captured in the tank. Hence, we could easily compare sources of widely differing spectra.

Denys Wilkinson, my thesis supervisor, had come up with a bright idea for a new type of source – using the photodisintegration of deuterium. What this boiled down to was, deuterium splits into a proton and a neutron, you can easily count the protons, so you know how many neutrons, too The actual source consisted of a small spherical glass vial of heavy water with ^{24}Na dissolved in it, producing neutrons by photodisintegration. We went to the Atomic Energy Research Establishment at Harwell to pick up the radio sodium which we were using as the source of gamma rays. This proved to be a problem since there was no means of transportation. My colleague Arnold Jones had just bought a Rover car, so in the end we used this. The car was novel in having a free wheel, like a bicycle, to conserve gas, so we would careen crazily down steep hills, with the engine idling.

Getting the radioactive sodium out of Harwell proved a problem. There was a guard at the gate, to prevent theft. It normally took several weeks to get a permit to take stuff out, and radio sodium has a short half-life, So we had to smuggle it out. Arnold would approach the guard and look suspicious, as though he were smuggling something out. The guard would search him vigorously. I would follow him looking nonchalant. It worked every time. I saw a detective story recently on television where the culprits smuggled radioactive material out of a facility – I was much amused

Unfortunately we had a puncture on the way back, so after stopping the car, I ran with the source some distance down the road, depositing it in the ditch, so we would not get irradiated changing the tire. Anyone seeing this would have thought we were mad. We had put lead blocks in the trunk of the car in front of the hot source. Anyone tailgating would have got what they deserved – a load of radiation!

In processing the ^{24}Na, some got on my hands, which drove the Geiger counter wild. It would not wash off, and for a week I slept with my hands over the covers, in order not to irradiate my vital parts. Fortunately, the half life was short. I left for Australia at the end of the experiment, and I always wondered how they got rid of the 500 gallons of permanganate. If it went down the sewer, it would kill all the bugs at the sewage farm, ruining the process. The experiment proved successful, so I was put on the International Neutron Calibration Board for a number of years (this was also partly because I had gone to Australia – and there were few university people Down-Under interested in neutron calibration).

One interesting experiment performed by a colleague to calibrate neutron sources in a large vat of water employed pellicles of nuclear emulsion (similar to photographic emulsion). It was found that condoms, because of their strength and thinness, formed the best protection for the emulsions. So this student had to go into the chemist's (pharmacist's) shop to buy one hundred condoms. The chemist was much puzzled.

3.15 TEA AT THE CAVENDISH

J.J. Thompson had left a bequest in his will to provide tea and cakes (cookies)

to members of the Cavendish Laboratory every day at 3:45 pm. The lady who provided the tea had the most immense metal pot – quite the largest I have ever seen. By the time I got to the Cavendish, graduate students were allowed only one cookie – the number of students having grown since J.J.'s day. Nevertheless, this presented the best possibility of meeting anyone you wanted to contact in the laboratory – everything stopped for tea, as the song has it, and everyone turned up. For the most part, the senior faculty sat together and talked, as did the graduate students and post docs.

I recall I sat for tea with Watson and Crick about the time they were working on the DNA molecule. They were then still junior faculty. What most people didn't realize was that their principal research was on the x-ray analysis of much simpler molecules. I advised them to quit working on a molecule which had thousands of radicals hanging off it, and stick to the simpler stuff for which they were being paid – no point in wasting their time. Of course I was a nuclear physicist, and apt to be totally wrong about these things! It was interesting having discussions with them. Crick had a loud voice and would discuss volubly about anything. Watson would sit quietly, and nod when he agreed with Crick. Yet he was the one who would write the excellent books on the double helix. Both became fellows of Churchill college, Crick left because Churchill accepted a bequest to build a chapel. It is interesting to note that at that time, molecular biology was still a small branch of physics.

3.16 THE SHEARWATER CAPER

My thesis supervisor Denys Wilkinson was very keen on birds. He was an avid bird-watcher, and had been for many years. He was particularly interested in the homing of birds. Did they have some internal magnetic system, for example? He had invented a device for determining how many hours birds spent in flight. It was a little tube, with a speck of radioactivity at one end, and a little photographic film at the other, and it was attached to the bird's leg. A ball-bearing in the tube shielded the film when the bird stood up, but when the bird tucked up its legs to fly the bearing rolled away, exposing the film. We always felt the birds probably walked home to fool him.

One day Denys invited me to help him release a bunch of shearwaters, seabirds from the north coast of Wales with immense wingspans – upwards of ten feet I believe. These had been sent us in cardboard boxes, to be released from a high point and followed using binoculars, to see if they took off immediately in the right direction. Cambridge is very flat, so the highest point was the library tower. We obtained permission, and went up the tower with the birds. They locked us in the elevator, so no-one could follow us, and I recall Denys wondering if our bones would be found there if anything went wrong. The roof at the top of the tower sloped away from us, so you couldn't see the ground directly beneath the tower. We started out releasing the birds by throwing them in the air, then following them out of sight. However, at the third or fourth, the bird simply fell over the edge of the tower, completely out of our vision. In our imagination, we saw a dead bird on the path at the foot of the tower. The next birds we threw up standing in the middle of the tower roof, and all went well, until one just flopped. We tried throwing it up several times, but always the same. The best thing seemed to be to put it back in the box, and return it. We saw no dead bird below the tower, and reasoned that it took off after falling several feet. This would not be surprising, because these birds must fall off the crest of a wave when taking off from the sea. I don't know what happens if it is calm. We took the bird back to the Cavendish, but that is not the end of the story. Somehow, it got out in the middle of the night, and scared the living daylights out of the porter (janitor) making his rounds. Can you imagine coming across a bird with a ten foot wingspan walking along a corridor in semidarkness?

3.17 The CAVENDISH MUSEUM.

A somewhat primitive museum had been established on the fourth floor of the Austin Wing of the Cavendish Lab. This was before the history of science was taken seriously. Nevertheless, the museum contained some of the most interesting artifacts. Maxwell's experiments on the viscosity of gases had been carried out in an inverted glass bell jar, supported by large cast iron feet that would have done better in a church. An even more interesting toy developed by Maxwell and on display was his Zoetrope machine. The slits in the rotating drum of the machine had been replaced by concave lenses of focal length such that it brought the image of the drawings on the far side of the drum to the center. The motion of the lens in one direction exactly compensated for the motion of the drawing in the other. so the image, lying at the center, did not move as the drum rotated. this produced a brighter and sharper image than the conventional slots. However, the most interesting feature was Maxwell's original drawings for this device, some of which were quite hilarious.

3.18 MOELWYN HUGHES – SWEARING IN WELSH

Moelwyn Hughes was my lab instructor for my physical chemistry course. We had to do permanganate titrations, where the end of the titration (running one solution from a burette into another) was indicated by the change from permanganate color to clear. I am color blind, and hence could not determine the end of the titration, so I would call over Moelwyn Hughes a native Welsh speaker, to tell me when the end came. This really irritated Moelwyn, not being colorblind, and he would swear at me in Welsh. It turns out that Welsh is an excellent language for swearing. Sir John Thomas, a Welsh chemistry professor from Cambridge, with whom I dined last night, agrees. Somehow it carries the feelings of the swearer particularly well. French, they say, is a good language in which to make love, but for swearing, Welsh is best.

3.19 COCKCROFT AND THE INTERVIEW

As my time at the Cavendish drew to a close, I was employed in the neutron calibration experiment. However, I had to look to the future, and it seemed as though I had three options – go into the army, go to America, or get involved in the atom bomb project at Aldermaston. However, Harwell had developed some fellowships, and there was a vague possibility I might get one, thus staying in England, and doing the sort of work I enjoyed. So I applied, and it ended up I had to have an interview at the Atomic Energy Research Establishment at Harwell. I went there with some trepidation to the most terrifying interview I ever had. Sat at a table were all the most formidable physicists in England, headed by John Cockcroft. I was placed at the foot of the table, so that all the participants could see and talk to me.

The interview progressed in normal fashion – questions to put me at my ease, (what books do you read?) followed by a discussion of my research. Somehow or other however, I got into a heated argument with Sir Brian Flowers about the nuclear barrier penetrability (how do alpha particles get into, and out of, the nucleus?). I had been interested in the physics of distorted nuclei, which was a new thing at the time, and I had shown this influences the penetrability. For some reason, Flowers violently disagreed with me, and an argument ensued. The fact that I was right did not matter – I was politically incorrect, and it meant that I did not get the fellowship, which meant in turn that I went to Australia instead – and for that I am forever grateful.

Still, I wonder what would have happened had I got the fellowship? Looking at my Cambridge colleagues, such as Geoff Dearnaley, I saw they progressed at Harwell – the most brilliant did very well. However, Harwell is now just another research institute, and England is no longer the principle power in the world scientifically and many researchers, such as Geoff, went into industry, on an international basis and did well. As for myself, although I was known in my research field, no one outside of it ever heard of me. Had I stayed in pure research, I would never have become involved in teaching, which ultimately proved to be what I enjoyed most. Many years later, I left the University of South Carolina to perform research at Yale. However, my desire to teach took me back to Carolina after a year to teach as well as do research, rather than just do research, as I would had I accepted an offer made to me at this time to go to Brookhaven Lab and work with Sam Lindenbaum and Luke Yuan, a very prestigious group (I believe Luke was grandson of the last Chinese emperor – but I digress – I love digressing!).

3.20 PARKINSON AND THE BEAUTY CONTEST

Prof. W. C. (Bill) Parkinson from Michigan State (who designed their cyclotron) worked with us on sabbatical. He had done much work on neutron cross sections during and after the war. He ate egg and chips at the pub with us for lunch (the cheapest meal). He and his wife decided on a tour of England, and they happened on a fair at one small village. Somehow, Parky got involved and ended up judging a beauty contest. I wondered about this until he explained. If a local had judged the contest, he would be an anathema to the losers ever after, but Parky, having judged, left and was never seen again – so he never had to endure the stigma.

3.21 PUBLICATIONS

It is interesting to see how stuff is published. My first publication, about the work I had done with Denys and John was published in the "Philosophical Magazine." How can a magazine be philosophical? Nodding its head with a long white beard. Anyway, at that time, there were sufficiently few nuclear physics experiments that this was the preferred publication. Next I published in "Phys Rev" (Physical Review) or "Phys Rev Letters" all of which were in the same issue. By then, the journal "Nuclear Physics" was running, so that is where I published next. However, my research interests changed, and also I became more interested in teaching – so "The Physics Teacher" seemed the obvious choice and I found it much more fun.

3.22 CAMBRIDGE (AND OTHER) CHARACTERS

INTRODUCTION

For some reason, to us students, the professors – I am using this in the American sense of lecturers – all appeared to be eccentric. I have since wondered whether I appear the same to my students, and have come to the conclusion that I probably do. So I have included notes on the impressions some famous (and not so famous) physicists made on me primarily at Cambridge

1 P.A.M. DIRAC

I attended Paul Adrian Maurice's course on quantum mechanics at Cambridge in 1950. I was a neophyte physicist then, mostly an experimentalist who was still rather weak on theory. Dirac's course was taken almost exclusively from his book, which at that time was still THE book on quantum mechanics, although being so symbolic, it was difficult for me to follow. Dirac had the appearance to me of being very old. Thin and stooped, he would lean over the narrow lectern he used until we were all convinced that both he and it would fall over. They never did, and the strange thing was as he got older he never changed. He came to Florida State University in the early 60s, largely, I believe, because his daughter's husband was a faculty member. I often wondered how he felt about life over here, he was so retiring. He used to take long walks in the country, which was probably difficult in Florida. I ran into him and his wife, at the airport in New York a year or so before he died, and I swear he looked just the same. I was on my way to Aarhus in Denmark, where he told me, his son had a position in the math department.

He was notoriously precise about his lectures. Weisskopf tells the famous joke that he (Weisskopf) said Dirac would answer questions after his lecture. One student said, "I don't understand that second equation professor Dirac." Dirac remained silent. "Aren't you going to answer the question?" said Weisskopf. "That was not a question, that was a statement," said Dirac. I took down his notes religiously, (which he wrote on the board) even though I frequently did not understand them. Dirac visited us here in the States many years later – I took out his notes, and now, miraculously, I understood them. Many tales are told of Dirac's absentmindedness. One I like – Dirac was married to Wigner's sister, who was pregnant. A student, seeing Dirac wandering about distractedly, said "What is the matter professor Dirac?" "Oh", said Dirac "Wigner's sister is having a baby". When he visited us I took him and Jesse Beams out for lunch. "Where shall we go?" I asked "Anywhere that doesn't have musak," stated Dirac. The two were immersed in the problem as to whether G changed with time. Their conclusion was that it was invariant using the evidence available then. I recorded an interview with Dirac for a short slot on an ETV news program I was involved in. Come time to broadcast, I said "I spoke with Professor Dirac about how he became a physicist". Immediately the broadcast cut to an interview with a famous black comedian – Dick Gregory I believe. The film chain had been set incorrectly, and could not be reset. So the camera went straight back to me, astounded at what had happened. I said "well, Dirac told me – etc.". This is the mark of a good television personality – never get phased.

2 SIR DENYS WILKINSON

Denys Wilkinson was my thesis supervisor. Tall and gangling, with trousers too short, he was rather a formidable person – but he had a very droll sense of humor. He was made fellow of Jesus college, and shortly after, he reported to us (who were his students), he noticed, while he was dining on high table one night, an animated discussion going on between the master of the college and a famous bishop. This went on for some time, until one of them said "Let's ask Wilkinson, he's a physicist, he would know." "Certainly," said the other. "Very well, if you were standing on the platform of Holborn underground station, and you peed on the live rail, would you get electrocuted?"

Denys vanished to Brookhaven National Lab in the summers, but we had our work cut out, so that was not a problem. We heard some rather surprising stories, not connected with physics about him, so when Dr. Patterson of Michigan State, who had spent a year with us at Cambridge, told us what had happened, it came as no great surprise. Apparently, Denys had gone to South Africa to work at the request of one of his students, Friedel Sellschop. Friedel had to go to the United States, and Denys had an affair with his wife, culminating in getting her pregnant. Friedel divorced and Denys married. What amazed me was subsequently, Denys got a knighthood.

3 G. I. TAYLOR

There was an active group working on the properties of ice at the Cavendish. J. W. Glen and J. Nye studied the flow of glaciers using a variety of blancmange. This peculiar British food is somewhat like aerated jello – with lots of other stuff thrown in. Anyway, if placed in a contour model valley, it will flow like the ice in a glacier – but faster, so it is ideal to study the behavior of real glaciers. The only problem was that, in the damp British climate, it had a tendency to grow mold, which gave the mock glacier a peculiar appearance after a while, not to mention a very odd smell. This was not quite as bad as the cockroaches which were kept in the old anatomy building. The ground floor of this building was occupied by the biology people, who were doing experiments on these cockroaches. The cockroaches lived in a wire enclosure and would eat anything – old boots, newspaper etc. They were under constant illumination, because cockroaches which are black in the dark become transparent in constant light, which is a great help in observing their insides. The top floor of this building was the prerogative of the cosmic ray physics people, who shielded their counters with large lumps of lead. They kept piling more and more lead on their equipment, until they came in one morning to find their apparatus in the basement rather than the top floor. The wooden floor had given way, and, of course, once it had gone through one floor it went through all the others.

To return, Glen, G. I. Taylor, myself and others would cycle for lunch from the Cavendish to the BR (British Restaurant) near the railway station. A relic of the war, one could get decent meals served cafeteria style, very cheaply – good wholesome food guaranteed to increase your cholesterol level by 30% – but just after the war, calories were a good thing – you needed calories to stay alive.

G.I. was a very knowledgeable, well-informed individual who enjoyed conversation. On thinking back he seemed very old to me, but I doubt he was older than I am now! He was short, white haired and had a sharp appearance.

4 G.I. AND LORD ROTHSCHILD

When I was running on the accelerator HT2, I well recall a dapper individual wearing a straw hat and carrying a cane (walking stick) passing my window every morning. This was Lord Rothschild, the English scion of the international family. He was an expert on bull sperm, and had, so I was told, written books on the subject. Somehow he got friendly with GI Taylor, who was taken with the fact that sperm can swim remarkably fast considering their size. GI was into hydrodynamics amongst other things, with the upshot that he made a gigantic model of a bull sperm, with Rothschild's help, and they towed this thing up and down one of the towing tanks used for ship models. The result was that they showed a sperm is ideally shaped for its purpose, perhaps one of the results of Darwin's theory.

5 MAX PERUTZ

Max Perutz occupied the room adjacent to me. He was working on the structure of hemoglobin, and had been doing so for several years when I got there. At that time Perutz was a very quiet, laid-back individual. I recall, when he was elected to be a fellow of the Royal Society, about the highest scientific honor you can get in England, he threw a tea party. Others would have had champagne at a big hotel, but this says a lot about Perutz. The Fourier analysis of X-ray diffraction pattern, necessary for the structure synthesis was exceedingly tedious. To do so, a bevy of girls was kept downstairs, using hand calculators – Friedens – to extract the coefficients. They would periodically get bored, and take a more exciting job – such as working the cash register at Woolworth's. Then more would have to be trained. Perutz had an incomplete model of the molecule, to which he periodically added an atom or two, as their positions became known. He finally completed the molecule after I left, and got the Nobel Prize – amazing – I would never have believed it! But genius, patience and intuition paid off. Two things speeded up his work – the fast Fourier transform, and putting one heavy atom in the molecule to shift the phase of the transform in one direction and avoid confusion.

6 NORMAN FEATHER

Norman Feather was famous for his natty appearance – never a hair out of place, all greased and brushed back. He was the examiner for the practical physics examination for the second and last part of the Natural Sciences Tripos which I took. Nobody had told us this, and I and my friends gave short shrift to this individual wandering round the lab and chatting to us as we were working on this very important experimental examination question. We thought he was one of the technicians. Particularly I had one friend whose equipment was in the window. We were using a ball bearing to measure the viscosity of glycerin, or something, which had a fourth power temperature dependence. As the sun came out and fell on his equipment, it would warm up, and he would get a low value. Then the sun would go in, and his values would suddenly increase – very distressing.

7 OTTO FRISCH

Otto Frisch, our prof, came to the Cavendish immediately after the war. He had been very active in experiments and calculations connected with the bomb during the war. As mentioned before, he coined the words "nuclear fission" having written the first paper on this, with his aunt.

I recall his recounting "twisting the dragon's tail", an experiment in nuclear criticality. A sphere of enriched uranium had been constructed, with a cylindrical hole drilled right through the middle. A cylindrical slug of enriched uranium was dropped through the hole, held vertically. Calculations showed the slug, as it passed through the center, should make the combination just pass criticality. The slug was dropped, and sure enough, criticality occurred as it went through the center. Unfortunately, the heat produced by the fission reaction made the slug expand and stick in the hole. This was a catastrophe – nobody wanted to go out and release the slug, because a potentially lethal dose of radiation would be received. The group waited tensely, trying to work out what to do. Luckily, after a short while, the heat produced also expanded the sphere enough that the slug was able to drop out. Everyone heaved a sigh of relief.

Frisch had the reputation of being less active at the Cavendish. He developed a pulse height analyzer which used a steel ball, rather like a pinball machine – the kick delivered by the electronic billiard cue impelled the ball along a tilted plane, and it fell into a different slot, depending on the impulse it received.

Frisch married an artist while I was there, who fed us pumpernickel, I recall (I had never had this before). He brought Lisa Meitner, his aunt around to see my research, and explained, in German, what I was doing. Why he explained in German I never understood – because she was quite fluent in English. She seemed a very nice old lady. He gave some quite good lectures on experimental nuclear techniques.

8 J.J. THOMPSON

J.J. Thompson was a very interesting man. I never met him – he was long dead – but several interesting tales survived. He came from a lower class background in the North Country, as opposed to the previous incumbents of the Cavendish chair, and was uncertain he could fill their shoes, but events proved differently – he was very successful – though it is difficult to imagine a time when he could pay his lecturers when he met them in the street with gold sovereigns. He was quite absent minded, and the tale I have is from a previous fellow of Trinity (Trendal) – his college. It appears his wife awoke one morning to find J.J. missing, but his pants over his chair. She immediately assumed he had walked out of college without his trousers on, being rather absent-minded. What to do? She called the porter at the main gate, in case J.J. had not yet got that far. Luckily, he was caught – wearing a pair of pants he had bought the previous day at a second hand sale – what we would call a yard sale – a thing he apparently loved. His son, G.P. Thompson, was master of Corpus Christi college when I was at Cambridge, and I saw him around the college – but he never came into the Cavendish, that I can recall, to take tea.

9 DOUGLAS HARTREE.

I received my first quantum mechanics lectures from Hartree, who gave the most simple graphic lectures. He never gave a lecture I did not understand – as opposed to Dirac whose lectures I had to ponder over. He employed the Schrodinger wave approach, and you could picture exactly what went on. He would come bustling into the class, red-faced – he had high blood pressure, and start off immediately. He was well known for his interest and knowledge of steam trains (as well as music – he conducted an amateur orchestra). The British railway system was suffering a lot of cutbacks just at that time, and he would frequently be one of those taking the last ride on a piece of track about to be closed. It was said he knew what every lever in the signal box at Liverpool Street (the station for Cambridge in London) did. He was writing his book on "Numerical analysis" where he said "Anyone intending to undertake a serious piece of calculation should realize that adequate checking against mistakes is an essential piece of any satisfactory numerical process. No-one, and no machine, is infallible, and it may fairly be said that the ideal aim is not to avoid mistakes entirely, but to find all mistakes that are made, and so free the work from any unidentified mistakes."

10 VICKY WEISSKOPF

Weisskopf became head of the Physics Department at MIT. In the early 50s he visited us to talk about nuclear physics – interesting lectures – after all he had just written "Blatt & Weisskopf", THE book on nuclear physics (Blatt had gone off to Australia, where he thought there was less chance of being hit by an atom bomb). Weisskpof's talk was in the Arts School, which housed the Math Department. After his talk he said Dirac had invited him over to hear what his students were doing. We went along too, because we felt maybe we would understand them better with Weisskopf there asking questions, since we never followed them otherwise. In fact, there proved to be a certain amount of truth in this. However, at one point one of Dirac's students demonstrated a problem, and Weisskopf said, "That's a very interesting explanation – who first worked it out?" The student replied, "You did." It turned out Weisskopf had written the article several years ago and forgotten it. He was very embarrassed.

11 G. F. C. SEARLE

Sir Lawrence Bragg on his retirement in 1954 gave a talk, "50 Years of Physics." I was sitting next to G. F. C. Searle, the grand old man at the time, who said to me in his cracked voice, "What's Bragg talking about today?" "Fifty Years of Physics," I said. "Ha," chortled Searle, "I could give one about Seventy Years of Physics," which he well could, since he was upward of 90 and still going strong.

I ran across Searle early on in my undergraduate career, running the experimental labs.

Later I was paid to help in the same undergraduate laboratory still presided over by Searle. Tony Hewish, who later got the Nobel prize in physics for discovering pulsars, worked with me. This was the prize where it was said Jocelyn Bell Burnell should have received it too. Hewish and Martin Ryle, who was also at the Cavendish as head of the radio-astronomy group, got the prize simultaneously in 1974.

Searle once invited me to look at his early research. The equipment dated from 1900, but worked well. For example, he recorded vibrations on a smoked plate, which was dipped in molten shellac to preserve it. He had retired before WW2, but had been brought back of necessity during the war to run the elementary labs which he had been doing since 1900 or so. He was notorious for criticizing the graduate student running the labs in front of the undergraduates – very embarrassing and a no-no. He also pinned a card on a non-too-bright undergraduate's back saying "Mediocre," and deliberately put a large magnet under the bench to produce weird readings of the earth's magnetic field.

As a child, he had known James Clerk Maxwell. (Searle was fifteen when Maxwell died in 1879). The rather forbidding portraits of Maxwell with his beard were very misleading, he said, because in fact, Maxwell was a very lighthearted chap, with a well-developed sense of humor (as is shown by his poem on "a telegram from a telegraph clerk (male) to a telegraph clerk (female)" given in the book "A Random Walk in Science"). Searle had been very sick in the first world war, and recovered with the help of the Christian Scientists, for whom subsequently he had a strong affinity. As mentioned earlier, he had assisted J.J. Thompson in many of his experiments, one of which was the determination of the Ohm. The apparatus was still in the Cavendish elementary lab, and each and every first year student had to make a determination of the Ohm – as I did when I was there. It worked by filling an enormous copper cylinder with water to find the volume. This formed the outer electrode of a cylindrical capacitor.

In one of his experiments, he had placed a mirror galvanometer in a student lab – but the meter scale on which the bright spot of light fell was missing one centimeter between the 20 and 30 cm marks – with the result that students plotting the voltage versus the current, say, had a jump in the plot instead of getting a smooth straight line. Many fudged their results, with the result that they got a "rocket" from Searle. As I mentioned previously, he also had a magnet hidden under the bench where we measured the earth's magnetic field. Again, the measured answer was OK – a fudged answer merited a strong reprimand. In America, I have often heard it said that such tricks are immoral – but are they? Or should we put up with students fudging their results, and us not being able to correct them? Too often, fudging of results occurs till late in life – and the culprits never get caught.

Cycling was Searle's principle exercise, and one Sunday he took off with my thesis supervisor, Sir Denys Wilkinson, on a trip through the local countryside. Noting a sermon was in progress in a church they were passing, Searle leaped off his bike, (followed closely by Denys). He entered the church, telling the minister in the pulpit, "Come out of there". The minister meekly came down, and Searle proceeded to preach a hellfire sermon. When he had finished, he descended from the pulpit. jumped on his bike (followed even more closely by Denys) and rode off. I later heard that this was not a totally unusual occurrence and that Searle was well known for his eccentricities.

What I had not known about Searle was that to some degree he had anticipated Einstein's special theory of relativity. Searle published a paper in 1897 in which he derived the expressions for the electromagnetic energy possessed by various moving charge distributions. This lead to an equation for a speed – dependent electromagnetic mass. [G.F.C Searle, The Phil. Mag., 1897, 340]

12 SEARLE AND HEAVISIDE

Strangely enough, he wrote a book about someone even more eccentric than he was — namely, Oliver Heaviside. Heaviside was a brilliant man when it came to the theory of current electricity — he invented Heaviside's operational calculus — but his personal life left much to be desired. After his wife died, he became an extreme recluse and even when he was made a fellow of the Royal Society — a very big deal in England — he never went to London to sign the membership book — so they had to admit him without doing this, a unique occurrence. Searle told me when he and his wife visited, Heaviside put a full half pound of tea in the pot — he had no idea how much was usual!

13 SIR LAWRENCE BRAGG

Bragg lectured me on optics. He had these yellowed notes at which he peered, but he gave excellent lectures in his rather high voice.

He felt that the undergraduate physics majors were too tightly focused on their subject — so he instituted a mid-term break — during which we worked all the harder. He also decided a history lecture might be a good idea, so he induced Herbert Butterfield to give us a weekly lecture on the history of science. This Butterfield eventually turned into a book, "The Origin of Modern Science." A very popular book — at the time there were very few on this topic.

Bragg of course, although he retired from the Cavendish professorship about 1954, did not retire from physics. He became director of the Royal Institution, the oldest independent research body in the world, founded by the renegade American, Benjamin Thompson, Count Rumford. It is perhaps interesting that the corresponding similar American institution, the Smithsonian, conversely was founded under curious circumstances as a result of the Englishman Smithson!

The movement of Bragg to the Royal Institution caused quite a brouhaha. It turned out that E.N. da C. Andrade, the previous occupant of the directorship, had not been told about Bragg's coming. He was a very contentious individual, and caused quite a fuss, ultimately having to be bought off, much to his glee. One of the nice tales about Andrade concerns growing large crystals. Andrade employed an old technician to grow his crystals, He came in one morning to find this individual drunk, sitting by the equipment — so he was fired and replaced by a bright young spark, who adjusted the temperature, and kept everything spick and span. But, he could not grow big crystals. Finally they got the old guy back — and everything went well again. It turned out that what the crystal wanted was to be left alone — even if the conditions were not perfect, one did much better not disturbing it. I don't know what the moral of this is.

Both Bragg and his father had the first name William. However, the younger Bragg always went by his second name Lawrence, and he used to get irritated by older people who called him William, in confusion with his father.

Unlike Dirac, who always thought in terms of equations, Bragg told me he had to picture something to understand it. This is perhaps why he was so good at the geometry of crystal structure. To improve his intuitive feel for crystal structure, he would play the game of three dimensional chess with his students.

Bragg lived out along the Backs, and rode in each morning by bike. He had an inordinately high, old sit-up-and-beg cycle, so as he travelled along, he was head and shoulders above the rest of the traffic, cars and such, and one could watch him approach.

He gave us tea once a year at his house, so us students could get to chat with him, and know him a little. He was quite a good conversationalist – unlike Dirac, whose comments were mostly "yes" or "no".

Steven Hawking also had a strange mode of transport, in that he rode a motorized wheel chair that seemed to travel as fast as a car. One time he zipped past us as we were sitting on the Mill dam. He whizzed by with his nurse chasing after him.

14 LINUS PAULING

I cannot avoid a note on Pauling. He had won two Nobel prizes, and was supposed to visit the Cavendish to speak – as I mention elsewhere, his son was there. This must have been about 1951 or so. Anyway, the US government took away his passport, and would not allow him out of the US because of the Communist fears generated by senator McCarthy This was the sort of action we in England were complaining about that the Soviets did, so to find what we had previously thought a civilized country doing the same thing did not improve our estimation of the U.S. A year later, his passport was restored, and he visited us. A very pleasant individual. Later, I attended a talk he gave at Los Alamos National Lab. It promoted vitamin C, about which he was bugs. Anyway, he lived to a ripe old age, so maybe he knew something.

15 DICK (RICHARD V DeR) WOOLEY

Dick Wooley was Commonwealth Astronomer (of Australia) when I met him. Living up at Mount Stromlo, he would ride down on his horse to the Australian National University and have dinner with us when he got bored. A very colorful character. One time I was out with him at night, and observed a very bright star. "What is that?" I said. "I dunno," said Wooley. "You'll have to ask an amateur astronomer!" He was related to Chris Wooley, with whom I used to ring church bells in England. Both had been to Cambridge, and, in fact Dick had run the telescope as a graduate student. He said, if they had visiting dignitaries, they would defocus the telescope a little. This had the effect of putting a brilliant colored cross in view, produced by diffraction effects around the mirror supports, the star being observed at the center.

He was at dinner with us when it was announced he had been appointed Astronomer Royal of Great Britain This gave him great glee, because his father, an admiral, had rather looked down on his son for becoming a sissy astronomer, rather than a macho navy man. Well, it turns out the Astronomer Royal is also an admiral, because of the connection with the eighteenth century longitude problem.

Unfortunately, on arrival in England he was asked what he felt of space research. He said it was all bunk, and a waste of money. The following week the Russians sent up their first Sputnik, and Wooley's name was mud. Some years later, he was staying at Cal Tech, as I was too, and he taught me to play a pool game called "cowboy". We were both very bad pool players. We both enjoyed folk dancing, and as Astronomer Royal he inherited Herstmonceux Castle, where, amongst other things, he organized folk dancing in the court yard. Many years later, Margaret Burbidge, a subsequent Astronomer Royal, confided in me she found supporting a castle to be an expensive nuisance, what with all the more conventional astronomical expenses.

There were many interesting characters at Mt. Stromlo. Alex Rodgers was then a student, a quite hilarious individual, who later became director. He married Ruth Dedman, who was a close friend of my wife Margaret. Then there was Walter Stibbs who came to work with Wooley on planetary atmospheres. He and his wife lived with us at University House.

16 ERNEST TITTERTON

I first met Titterton in 1954, when he came through Cambridge on a recruiting expedition for the Australian National University. I was an ideal candidate, finishing up my PhD in a field he was promoting. He was always an overpowering personality, with a loud, cheery, penetrating voice. In Australia, he ran the nuclear physics program at the Physical Sciences Center. He was an aggressive leader, but sometimes carried his aggression too far. One time I needed a large – very large – sodium iodide crystal to measure the energy of gamma rays, and charged him with getting one. He immediately said how much would it cost, we did not have the money, and so forth. Yet two weeks later he came back and said he had raised the funds, and I could get the crystal, which in fact proved a valuable investment at the time. He had worked at Los Alamos on the first atom bomb, (he was well known there as a great jazz pianist) and had been the person charged with counting down when the first bomb was exploded. They took a movie of this event, but when it was about to be released, noted the guy counting down had a British accent. So they dubbed in an American voice. This vastly amused Titterton. He had a running fight in the press with Harry Messel of Sydney University about the dangers of nuclear testing. Messel pointed out that the effects of nuclear radiation reduced fertility, so the bombs let off in Australia meant that fewer people would be born – over the years maybe 1000. So it was as if these had been killed. Titterton immediately responded by saying that the increase of radiation would be the same as if we lived up hill 1000 feet high. Both were right.

Many years after I left he had a bad accident, and became a quadriplegic. He wished to die, but survived many years.

17 ROBERT OPPENHEIMER

I met Oppenheimer in the sixties, just one time at a physical society meeting. He was much shorter than I had thought, and he didn't say very much. However, his reputation is well known. What is perhaps less well-known is that he went to the Cavendish in the thirties as an experimentalist. Mark Oliphant told me Rutherford came in to find Oppie banging on the floor with a hammer. It turned out, his experiment was not working, and rather than attack the equipment, he took out his frustration by beating on the floor with a hammer. After that Rutherford made him a theoretician.

18 H.N.V. TEMPERLEY

From what I have said, it might be thought that the Cavendish had a roster of brilliant lecturers. Nothing could be farther from the truth, because being a brilliant researcher does not make a good lecturer. I recall hearing Charles Townes here in South Carolina give the dullest seminar shortly after winning the Nobel Prize. He later improved dramatically, I may say.

One such Cavendish lecturer was Temperley. He gave a series on solid state physics, and started out with ten or so students. After the fourth lecture, he was down to me and another guy. He had an unfortunate stutter, amongst other things. As with many stutterers, you knew what he was going to say, but he could not get it out. I found this most frustrating.

19 WATSON AND CRICK

It is perhaps interesting to note that we all had small laboratory rooms in the Austin wing of the Cavendish Laboratory. In the lab on one side of me were Watson and Crick, and on the other side Max Perutz. Molecular biology was a small branch of physics then, with everyone knowing everyone else in the field. Hence, I saw quite a lot of them apart from having tea. The synthesis of organic molecules was a very tedious business without computers, incomprehensible to today's students. It involved much physical intuition, and guesswork. If you got the right answer it was much less difficult to confirm it. The culmination of it today with the synthesis of DNA would have been unbelievable to the students of that era. The meaning and significance of its synthesis provided us with much tea time discussion.

4 ENGLAND TO AUSTRALIA

4.01 LEAVING OLD ENGLAND

As my graduate student tenure started to run out, I wondered what to do. I particularly wanted to spend some time in the United States – not permanently (which finally occurred), but for a short period, as many postdocs did. So I made enquiries and applications, and was accepted at Brown University in Rhode Island. Then I hit a snag. I had not fulfilled my National Service – which most students performed much earlier, but I had been deferred on account of my research. I was not anxious to be conscripted, because about this time most recruits were sent to Israel, where they spent their time separating Jews from Arabs and vice versa. It struck me that nothing could be a more unrewarding (and dangerous) activity than this. At that time, all the Arabs and Jews wanted to do was fight one another, and to stand between them and berate both sides was not profitable (and could be lethal – the Scots have a saying "beware the redding strake," meaning, don't try to separate a fighting couple.) The alternative was to work at Aldermaston, the Atomic Weapons Research Establishment, on the atom bomb – which also did not attract me. The National Service would fall due before I could take up my position at Brown. Then came the solution – and a perfect one at that. Prof. Ernest Titterton came around from the Australian National University (ANU). The university itself, in Canberra, the capital of Australia, was new, and the physics department even newer. My colleague in research, John Carver, had just gone back there, as had the technician (Hawkins) who made my equipment – and I was offered a place as a research fellow, to take out to Australia, install, and operate the Harwell synchrotron, a 35 MeV electron machine which had been donated by the British government to ANU. No one in their right mind could resist an offer like that – and I would have to leave almost immediately, thus legally avoiding my national service. Of course I accepted, and went to Harwell to dismantle the machine. I stayed at the Horse and Jockey Inn, near Harwell, a pub famous for the exceedingly attractive daughters of the publican, who acted as barmaids.

4.02 HARWELL

Harwell had been founded on a defunct RAF airport, used originally for bombing Germany. The large hangars proved ideal for building big pieces of equipment covered with shielding and such. The synchrotron was in one such building, and we proceeded to take the machine apart. I was given a very old technician to help, and he insisted on labelling every little piece of the machine as it was removed. Impatiently, I wanted to simply load the stuff ready to go. It wasn't until I tried to put it back together in Canberra that I thanked my lucky stars I had had this guy. As it was, there were pieces of the machine we never did find where they went. At that time, much in Harwell was secret, so we were searched going in and coming out. Anyway, I had had problems with this in the past, because we needed to take radioactive sources back to Cambridge, and the bureaucracy involved in getting them out was prohibitive. The standard practice was to go out with a friend, who would deliberately look suspicious. The guards would search him, and, being satisfied, if you were lucky, leave you alone. Nevertheless, although we were never caught, I realized I would never make a good spy. Anyway, after a couple of weeks, the machine was dismantled and packed, ready for shipping to Australia. Because it was going by freight steamer, it would arrive long after I did.

So back to Cambridge I went, to wind up the miserable experiment I was doing on calibrating neutron sources at the old Short Flying Boat factory on the Madingley road, where I had the 500 gallon tank of permanganate solution that I mentioned before.

The Short plant was also occupied by the people developing the projected linear accelerator (similar to the big, two mile long Stanford machine) The staff had spent years seeing how they could reduce the cost, a big mistake because it was finally cancelled. You should always spend the money you ae given! The staff all went to the new CERN (European Nuclear Research Institute) and did very well, much to their surprise. A joint European consortium such as this had never before been successful.

4.03 AUSTRALIA BY SEA

For a kid who had never even left England, the idea of travelling half way round the world was awe inspiring! Since this is a way of life that has completely vanished, I shall describe the passage to Australia by liner in some detail. It normally took a month to go out via Suez, even on the fastest boats. They differed much in design from those that plied the Atlantic, in having many open decks, since most of the time was spent in the tropics. Atlantic boats were enclosed because of the vile weather.

My parents and I travelled by rail to London, and I visited Westminster Abbey for the first time and, incidentally, met an old school-friend by the name of Bullough (who later became a minister I believe) in the abbey. The boat train took us to Tilbury, where the old Orion lay at dock. A member of the Orient line, she had been in use since well before world war 2, and was a typical member of her class. All the boats had names starting with O – Otranto, etc., and were not very large by modern standards – holding, I suppose a couple of hundred passengers. She did not even have air conditioning, except for the dining room, so we would stuff ourselves in the cool air, go on the hot deck and regret having eaten so much. The cabins were cooled by a system known as the "Punkah-louvre". This seemed to consist of gigantic scoop devices on deck facing the direction of motion, which diverted some air into the cabins. All the Orient boats were a delightful cream color.

I had a first class berth, (the only time I ever sailed first class), which I shared with a young guy of my own age who was a mechanic – at that time Australia badly needed trained technicians. My parents had thoughtfully bought me evening dress – dinner jacket (tuxedo) etc., because their idea of first class was that we dressed every night – much like "Upstairs/Downstairs" on television. In fact, only on rare special occasions did we dress – the ideas of the middle class about the upper classes are somewhat mistaken. The only other thing I bought of any value was a Voigtlander 35 mm camera. This I got tax free – it was waiting for me on the boat and it proved the most useful thing I could have got – the Kodachrome slides I still value. Two British lines took most of the traffic to Australia. They both went through the Suez canal, but whereas the Orient boats stopped at Colombo in Ceylon (now Sri Lanka) the P & O boats (Peninsular and Orient Line – the peninsular refers to Spain) stopped at Bombay (now Mumbai). Since my passage was paid by the Australian government, I would normally not have travelled first class, but all the boats were booked up, and this was the only passage available. I did not complain, naturally, and it proved superb, because rationing in England finished the very day I left (I still have my ration book), and on the boat we had gourmet food from all over the world. The steward would allow us to go through the menu twice, yet I did not put on weight – because in England I had been eating bread and potatoes, and on the trip, fish and steak.

Ah, those were the days! Travel by boat was most common then, because air travel was noisy, tedious, and uncomfortable – only slow, relatively short range, propeller driven aircraft were available. By sea, you travelled luxuriously in a comfortable berth, attended by servants, and eating the best food. I have never lived as well since. Unfortunately it is a vanished way of life – only the cruise ships give the slightest vestige of what it was like – and there it is only for a week. In a month, you got to know most of the passengers quite intimately. I was asked, didn't you get bored? Never once. There was much to read, they had movies and dances – dancing on a ship making way in a rough sea is quite an adventure. The dance floor would tilt in one direction, and everyone would dance downhill – then it would reverse roll the other way and we would all dance back downhill again. The word "posh" meaning stylish or expensive, arose from the fact that, going to India, the best berths were Port Out, Starboard Home.

It was sad saying goodbye to my parents, although they knew I would be coming back. Little did they know I would return for only a week or two in four years' time, and then only to leave for America.

We sailed from Tilbury on the tide, making way slowly down the Thames and out into the Channel. The Bay of Biscay is notoriously rough – so I prepared for sea sickness. A fellow passenger recommended créme de menthe as a cure. Believe me nothing could be worse – nevertheless, I was not sick then, nor during the whole passage to Australia.

4.04 GIBRALTAR

Our first port of call was Gibraltar, a British possession since the time of Napoleon, (much to Spain's chagrin) and a very interesting place. At each port of call we would have one or two days, while passengers were taken on or dropped off, and provisions and fuel loaded on board. This was pleasant, for it gave a few days anticipation before our next port. The Rock itself was like a Swiss cheese, hollowed out to store water, and other things in its defense. At that time, relations with Spain were relatively good, because the town provided much employment for people from across the border, who crossed no-mans-land every day to work. Gibraltar was a free port with no taxes, so almost every passenger ship entering the Mediterranean would stop there. We visited the fort on top of the Rock, to see the famous apes – ostensibly when these leave, so do the British. You could see across the small airstrip, and the barren no-mans-land, into Spain, on the one hand, and across the Straits to the Atlas mountains in Africa on the other.

4.05 NAPLES, POMPEI, CAPRI, ETNA

Our second port of call was Naples – "see Naples and die" is an old saying – but some people said that after seeing Naples they felt like dying. We were driven by bus (a curious bus – although they drive on the right in Italy, the steering wheel was also on the right) through the slums of Naples, with the washing hung on lines running from an apartment on one side of the road across to the other – at the sixth or seventh floor level. We went to a factory making brooches and such of little tiny pieces – cameo, mosaic, and intaglio brooches of people's heads. Then on to Pompei with the volcano Vesuvius in the background. Herculaneum was still, at that time, not open, but Pompei was a most interesting sight – the tracks in the stone of the chariot wheels. However, the thing that mostly comes to mind was a visit paid to the house of the Vetii – two bachelor brothers who ran what appeared to be a cross between a bachelor apartment and a brothel. Anyway, there were a number of very explicit sexual paintings – murals – on the walls, which were very interesting to a naive Englishman, such as I. The authorities would not allow women to see these paintings, so all the men trooped in – and you can guess what happened – as soon as we came out, the women wanted to know what it was that they had missed. Then back to the boat.

We sailed past the Isle of Capri, where you could see the blue grotto at great distance, and the two parts of the island – Capri, and Anacapri. Then through the straits of Messina, where it was dark enough to see the volcano (Etna) – the red hot lava in the crater throwing light on the clouds above.

Our next stop was at Kalamata, on the Peloponese of Greece. This was right after the Greek earthquake, which had destroyed completely the homes and livelihood of many Greeks – so they had decided to emigrate to Australia. The Australian government would, I believe, pay their passage out, but only the cheapest way. It was expensive to anchor in the Pireas – the port of Athens – so we anchored off Kalamata, and the emigrants came out on a large raft – it was one of the most moving experiences of my life. Here was this magnificent liner (for the time) white, illuminated, and with flags flying, and these poor emigrants came out with all their belongings – and all their families on this raft. The raft tied to the liner, and as I looked down, the families wept, kissed and parted – perhaps never to meet again, travelling half way round the world.

4.06 PORT SAID

Next stop the Suez canal. This was before all the trouble – so we anchored at Port Said, and the bumboats came out to sell us whatever they could. Believe me, black velvet pictures are not a new invention. What completely confused me was that the women passengers were all referred to as Mrs. Simpson. This was no doubt the wife of Edward VIII – but why, long after WW2, this should still have persisted is beyond me. The British Royal family is evidently of world-wide interest. On the dock I saw a briefcase I liked – but clearly, the guy was asking too much. I am hopeless at bargaining, but my cabin-mate was good. "Leave it to me," he said. After a while, he had got the price down to a reasonable value. "Now we walk away," he said. So we did. We had gone five yards when the purveyor said, "OK, I give you a good price," and cut it in two. I used that briefcase for the next ten years – and as I looked at it, it reminded me – I am no good at bargaining. I saw two men hand in hand – obviously enamored of one another. The same thing I am sure happened in England – but there it was much more circumspect. It made me think.

At the end of the Suez canal there was a large statue to the Australians who had lost their lives at Gallipoli, in WWI. This was later torn down during the troubles over the canal – which was one of the items that really irritated the Australians.

We sailed down the Red Sea, and it was the hottest I have ever been. The cabins, not being air conditioned, were unbearable, so we decided to sleep out on the deck, and did so. When we got up in the morning we could have performed in a minstrel show – we were completely black. What we had not realized was that the smuts from the smoke from the funnel would fall on us since we were behind the funnel. Evidently as the night wore on we got blacker and blacker. We only did that once!

Aden lies at the end of the Red Sea, and is part of Yemen where the recent troubles culminated. We approached past the clock tower to our anchorage. There is an extinct volcano behind the city, and many people live there. The place seemed to consist mostly of sand and volcanic ash, so it is most unattractive. Nevertheless, coming from Europe, it seemed quite exotic. We were in port for a relatively short time, since the purpose was to fuel the ship with the Gulf oil and be on our way.

Again, it was a free port, and I bought an Aldis slide projector (manufactured in England). It has proved useful over the years, but now computers have replaced it. The shops all opened directly onto the street, wide open in front, quite different from Europe, so it was easy to see what they were selling.

4.07 CEYLON (now SRI LANKA)

Our next stop was Ceylon – Sri Lanka. There I met an old friend from Cambridge, George Disanaika. Oddly enough, he was later to precede me to Columbia SC, and spend a year there ahead of me. Colombo was a nice place – it was my first really tropical port of call, and I well remember my introduction to the university there. It was exceedingly hot and humid. I found George working away at his desk – but this sat under a thatched roof – no walls, with a tropical fan overhead to provide a draft. I believe it was just before the cooling influence of the monsoon. We visited the zoo – where I saw a family of elephants, each holding on to the tail of the one ahead. And a sign in three languages – Tamil, Ceylonese, and English: "These animals are dangerous." As we drove through Colombo, George would point out, "And this is where the mistress of the prime minister lives." I wondered how many mistresses he had?

When at Cambridge, I had a friend whose father had the profession of writing constitutions. Each old colony of Great Britain, as it became self-governing, was provided with a constitution, and someone had to write them. So this individual went out and discussed with the locals how the constitution should read. It is to me, now, most interesting to see what failures many of these have been. Ceylon was certainly no prize, and Uganda – Ugh – Kenya, better – but still? Why is this? It is clear the fault lies in the people rather than the constitution – any constitution can be subverted. George Orwell saw this in "Animal Farm" – "All animals are equal–but some animals are more equal than others." Anyway, I liked the king coconut milk or juice in Ceylon.

4.08 COCOS KEELING ISLANDS

Then on to Perth, via the Cocos Keeling Islands, which lie in the middle of the Indian Ocean. They are small and very isolated. As we passed them close by, the islanders put out a small boat, and threw out a container with the mail being sent, which was picked up by the Orion (after it hove to) which, in turn, threw out a container of mail for the islanders.

The king of these islands, for many years was a member of the Clunies Ross family, who went out from England to these little islands, and decided to become king. I met his relative, Sir Ian Clunies Ross in Australia, head of CSIRO, the research institute of the commonwealth, and famous just as I got there for having been inoculated with the mixamatosis virus, used to kill off the rabbits. People were afraid it would kill them off too, so Clunies Ross, and the woman who had discovered the virus were inoculated, and unaffected. During the period the virus was active, they were able to run twice (or more) as many sheep or cattle on the same property – because there were so few rabbits.

4.09 PERTH

I found Australia fascinating. We first arrived at the port of Fremantle and travelled by rail to Perth. What struck me most was, we had come half way round the world, through strange and exotic places, yet the houses and buildings in Perth were just the same as those I had left in England. The one novel feature was a large car park outside the University of Western Australia at Perth. That would never have happened in England. The flowers, too were different, and very colorful.

We rounded the corner of Australia and entered the Great Australian Bight – to be hit by the most humongous storm. The waves in the bight can be immense – much larger than in the northern hemisphere, because they travel completely round the globe in building up. The result was, the ship rocked through angles of greater than 30° – I have pictures to prove it. At first the waiters damp the table cloths at dinner, then they raise ledges around the edge of the table, to stop the crockery falling off, and finally, they tied the furniture down – the last actually happened to us – and it is a curious sensation when going to sleep under these circumstances, because one minute you float above the bed, as the ship rocks down, the next it is pushing up on you with twice your weight. Practically nobody got sick. We had been on the boat at least three weeks, and were used to the motion.

4.10 ADELAIDE

Our next stop was Adelaide. We disembarked at Port Adelaide. Adelaide proved to be a pretty, very quiet place – laid out on a rectangular plan, unlike towns in England, but very like America. Mount Lofty ranges there were very attractive. My cousin finally settled in that region, and he and his family love it. We visited them some forty-five years later.

4.11 MELBOURNE

Next stop, Melbourne. We took the train from Port Melbourne to the city, where I had the most bizarre experience. There was a bar directly opposite the railway station in Melbourne, and it opened onto the sidewalk. As we came out of the railway station, they were rolling drunks onto the sidewalk, so that they could pull down the shutters on the bar. This was apparently a normal experience at 5:30 or 6:00 pm whenever they shut up. I was baffled, but it turned out that in their wisdom, the government had decreed all bars must shut at 5:30, so the workers would go home to their wives and families. The result was, these people, knowing there was no more booze that night, arrived there at five, and simply soaked up as much as possible in the remaining half hour. I gather the situation is no longer the same, but it shows how laws, passed with the best of intentions, can go sadly awry. I have little recollection of Melbourne, other than the trolley cars which were still running there, (and also at my revisit in the late nineties when a small earthquake hit us while on the tram).

4.12 SYDNEY

Our trip to Sydney was beset by one problem – noise and vibration. We had lost time during the storm in the Bight, and to make it up we went full steam ahead. The whole ship juddered and shook as if it were about to fall apart, with the engineers trying to get all they could out of the engines. This stopped one getting a full night's rest. Anyway, we entered Sydney harbor, passed under the bridge, and anchored around the point from Circular Quay, the heart of Sydney. I flew from Sydney to Canberra – my first flight, in what I took to be the acme of perfection in aircraft – a DC3 (C45, or Dakota). This seemed to me to be a marvelous plane, convenient, comfortable – and it got me to Canberra in no time at all. A few years ago, I flew to Atlanta on a similar plane. In the intervening years it had become uncomfortable, it rattled and shook, the engines were very noisy – and of course, it was not pressurized or air-conditioned. Amazing how our views change with time!

5 CANBERRA

Canberra in 1954 was little more than a village – just under 30,000 inhabitants – all legislators, embassy people, or associated with the university, plunked down in the outback at 2000 ft. altitude surrounded by small hills (which periodically had bush fires – seeing Black Mountain burn, lit up like a Christmas tree, was quite an experience). The tales told of its early days were quite bizarre – like, when the prince of Wales came to open or inaugurate the city in 1929, or whenever, there were no trees – but plenty of forests nearby. So a day or two before the prince came, they chopped lots of trees down, and stuck them in holes down the avenues. For the opening, it looked beautiful – but a week later they all drooped and died.

5.02 UNIVERSITY HOUSE (AND SKIING)

Arriving in Canberra, I was put up at University House, a brand new building, palatial by my standards – and by Australian standards too. I moved into a new apartment on a cold day, so turned up the thermostat – nothing happened – so I turned it up some more – still nothing. Finally I turned it full on – great disappointment, in fact the apartment got cooler! At dinner that night, it turned out I sat next to the guy whose apartment was directly above mine. "You know" he said, "a weird thing happened to me today. My room started to get hot, so I turned down the thermostat, but the more I turned it down, the hotter it got!" Well. of course we worked it out. His thermostat was hooked to my heating system and vice versa. That happens in a new building in Australia I suppose!

University House was run much as a traditional English University, and governed by a Master, a man by the name of Trendall, who was an expert on Greek vases, among other things. I recall one time at dinner, a servant arriving to say a communication had come from the Pope that Trendall had received an award – Knight of the Virgin Fleece, I believe, which we felt very appropriate, since Trendall was unmarried. The cataloguing of the Papal pots involved Trendall in much traveling. He came up to me one day with a newspaper, most upset. It said, on average, there was one plane accident every, lets say, 250 thousand miles. "I have just worked out," said Trendall, "That I have been 250 thousand miles. Should I quit flying?" It took some effort on my part to convince him that the probability of an accident was just the same as if he had never flown before.

The permanent residents of University House were quite varied. An English law professor called Sam Stoljer took up residence directly above me. He was a short, humorous individual. He took up with a student named Dagmar, and I recall the lady who cleaned our room complaining she found Dagmar's stockings drying in Sam's bathroom. Anyway, I recall the Master complaining that he did not mind this relationship existing, but the trouble was that they made it so blatant. So, Sam had to go, and eventually lived in a house rented out by the university, nearby. Another individual was a former Rhodes scholar, Bob Hawke. He and his wife returned from Oxford on the same ship as my wife. We used to play cricket, and in fact, he attended my wedding, primarily, I think, because there was nothing else on. Many years later – about thirty or forty, I saw his picture in a paper, and was astounded to find he was prime minister. He was into labor law, and Menzies, the conservative prime minister, was in power when I lived in Australia. It seemed impossible then that Hawke could ever amount to anything.

In the first week I was resident in Canberra, I was told that there was a group going skiing, and would I go along? I said I would love to, but my clothes had not yet arrived from Sydney. So it was I went skiing in my best (and only) double breasted suite. I have a picture to prove it. The YMCA had an antiquated bus they took up to mount Kosciusco. Unfortunately, on the way up the engine fell out adjacent to the old hotel which had burned some years earlier. We all had sleeping bags, so we camped out in one of the burned out rooms, sliding together in our sleeping bags, like sardines in a can. The following day we managed to go on. Smiggin's hole was our venue. Things were very primitive in those days. A farmer had put an old car at the foot of a hill and jacked up the rear end. A rope went over the rim of the back wheel after the tire was removed, up the hill, around an an old wheel at the top, and back down to make a loop forming the rope tow. I had never skied before and there was no instruction – it was every man (or woman) for himself (herself).

We did a lot of cross country skiing, and one time we skied round a mountain, which got steeper and steeper as we went – but we didn't know how to turn round to go back. Unlike the alps, most Australian mountains don't have a sharp drop off, so if all else fails, you sit on your skis and slide down. In spring, it is warm enough to ski naked to the waist. We now realize, this promotes cancer – but we did not know then. However, we got a lovely tan, which, for some reason, faded very rapidly. Some said it was because the light reflected from the snow could not give a long lasting tan.

5.02.1 WOMBATS, LIGHT BULB SOCKETS, AND TICKLING TROUT

One of the more interesting marsupials common to Australia is the wombat. They are about the size of a large hare. I had the unfortunate experience of meeting with one of these creatures in the middle of the road on the way driving to the Australian Alps. Neither of us was fast enough to avoid the other (wombats are rather slow) so I ran over its head. Much to my surprise, the animal got up and ran off as though nothing had happened. I believe they must have a very tough skull. Perhaps that is why they have survived so long.

Two amusing incidents involved the reflecting pool in front of the society rooms at University House. I had just brought a new colleague Bill Turchinetz and his wife to the House. It was dark and I was taking them to their rooms. The pool, only a foot or two deep, abutted the walkway, with the result that Bill, not seeing it walked right in. He did not fall, but waded to and fro wondering what had happened. Some days later he appeared in our doorway rather frazzled. It turned out that the light had burned out in their bedroom. He was from Canada, so naturally he unscrewed the light bulb, twisting it round and round as one would with an Edison socket. Suddenly there was an immense flash, and all the lights went out. What he had not realized was that Australia, in common with England, used the "bayonet" socket, so instead of unscrewing the bulb, he had been twisting the wires together, until they finally shorted.

The pond had large goldfish, or small carp in it, and I had just read an article on tickling trout, so decided to try it in the pond. The streams in Scotland have overhanging banks, beneath which the trout like to lurk, swimming slowly against the current to stay in the same spot. The fisherman lies down on the overhanging bank, unseen by the fish, and dangles his hand in the water just behind the fish's tail. Gradually moving his fingers forward he tickles the underside of the trout (which evidently they like) until the fingers are right behind the gills. He can then flip the unsuspecting trout onto the bank. I tried this out in our pool (much to the surprise of passers-by) and, in fact was able to catch the fish, though it wriggled out before I got it on the bank.

Much later at Oak Ridge in Tennessee, I could not convince my colleague that this was in fact a real method of catching trout. I won a bet. He called the state wildlife service and asked, "Is it legal to tickle trout in Tennessee?" Much to his surprise, the wildlife person, instead of laughing, replied that it was quite legal, but you had to have a valid fishing license.

5.03 ANTHROPOLOGY

Because of its situation bordering the Pacific, the ANU had a very strong anthropology department. I recall Dr. Stanner, a well-known anthropologist, holding forth on the book, "Coming of Age in Samoa", by Margaret Meade, which he thought a fake in that the author had been hoodwinked by the natives, or worse, made up much of the book herself. Marie Reay and other anthropologists often went on our picnics, and regaled us by deflating the myths about Polynesia, Micronesia, etc. The sexual proclivities of the natives was quite interesting – being for the most part much less inhibited than ours. Oskar Spate was also around.

Then there were Kath Jupp and Norma McArthur in my wife Margaret's demography department (often mistaken for democracy) and Charles Price, who first convinced us dried milk was better than natural. We would picnic on the Cotter River, as a group. We saw a snake swimming with us one time, and the argument arose, can a snake bite underwater?

5.04 HIGH TABLE

As a new university, ANU tried to absorb the traditions of some of the older institutions. Professor Trendall, having been a fellow of Trinity College Cambridge, thought we should establish a tradition of high table, a low dais at the far end of the dining hall on which the "dons" (people who ran the residential college) sat. I must confess I have always rather liked this, but it does smell of privilege – something Australians are very much against – as am I too – my father having started out as a laborer in a tan-yard. Anyway, come dinner time in the hall, we all put on our gowns, and marched down the aisle to the high table. The master sat at the head of the table, and one of the fellows read grace. It was rather a delight to have grace read in Latin, pronounced with a broad Australian accent. Dr. Joplin was famous for her peculiar intonation.

If you wanted wine, you merely signed your name, and for a paltry sum, you could consume as much wine as you wanted. Even then, the local wine was very good indeed – I myself feel there is none better in this world. Since we were entitled to drink as much as we wanted, it was very tempting to consume a little too much, and then the the problem was not to reveal this as you tottered from the table, proceeding to the "combination room" or drawing room, where we would have brandy and cigars – ah, what a life – and long playing records had just come in, so a nice Schubert or Beethoven Quartet? Anyway I have never lived as well – or got as much work done! I discovered several of the physicists, retired from the university, had developed vineyards nearby.

5.04.1 THE PEOPLE AT HIGH TABLE

In England, the Dons sit at high table, so in Canberra it was the heads of departments and such dined there. One such was Sir John Eccles, then simply Jack Eccles.

5.04.2 JACK ECCLES

Jack Eccles ultimately won the Nobel Prize – not in physics however, although visiting his lab with Rose, his daughter, I found it had more electronics than our physics lab.

He had certain rather curious ideas. Being Catholic, he had numerous children – nine I believe, and he felt each child should have their own room. So arriving in Canberra, he had such a house built, each child having their own room. His wife was a rather unusual person too. She had a quiet hour in the middle of the day. As you can imagine, with nine kids life was somewhat chaotic, but at the set hour, she would go to her room and quietly sit down and meditate. Rose was exceedingly bright, and I used to play tennis with her.

5.04.3 C.P.(Patrick) FITZGERALD

Patrick Fitzgerald was one of the most interesting people on high table. He was an old China hand – in fact he was the old China hand, having been everywhere in China before the second world war. I have a vague feeling he was in some way connected with the railway company. Anyway, he had fled the communists, and somehow escaped under fascinating circumstances. Fitzgerald would get into great arguments with the professor of Chinese history at Canberra University college, who also dined on high table. Fitzgerald believed that communism in China would ultimately change, or be changed, because of the very deep familial ties the Chinese develop. I have a feeling that this is true, but we have yet to see a complete change, and certainly not a reversal.

The tale Fitzgerald liked to tell was of workmen sent to change the sign over the great gate into the forbidden city in Beijing, which represents the ruling dynasty. Having removed the sign, the workman said to his boss "what shall I do with it". The boss thinking, well you never know, the regime might fail, said "put it in the room over the gate". The workman did so, but reappeared with another sign, saying "look what I found here!". It was the sign of the previous ruling dynasty, many hundreds of years old. Someone of an earlier generation had thought the same thing.

5.05 THE WINE TASTING COMMITTEE

The wine tasting committee is the best university committee I have ever been on. University House had a vast cellar under the main hall, and anyone seeing it said what a fine wine cellar it would make. So somehow, we convinced the university of this, and put together a committee whose sole purpose was to buy wine, and lay it down in the cellar for sale later to the students. It was my very good fortune to be placed on this committee. Most university committees are acrimonious and argumentative in the extreme. This is the only committee where harmony inevitably prevailed. We would send away (on university stationary) to all the vintners in Australia and request bottles of their best wines, as samples to decide whether we would buy a barrel or two. Many is the happy Tuesday evening we would spend tasting these wines, then we would order a barrel, and bottle the wine. We had a concrete plinth installed in the basement for the barrel, and we would tap it into bottles sterilized in the autoclaves of the medical school. One time Baron Fazekas de St Groth, a congenial member of the medical school, had the misfortune to drop a barrel and break his leg – he had quite a time explaining that. My own forte was fortified wines such as port – and believe me, there were some really excellent port wines available in Australia. So much so that after a year or two, we decided to put a few bottles on sale. Now, the port wine was supposed to take seven years to mature, but it was so good that before you could turn around we had sold the lot! I believe this committee is still in existence, but it probably takes influence to get on to it now!

One winter vacation we had a conference of psychiatrists in residence. I had stayed on, and believe me it was rather an odd experience dining with these characters. I got the distinct impression that they were analyzing one another all the time. Of course that could just be me. Trendall got a call late at night from Sydney that one of the doctor's patients was playing up – threatening suicide, etc. We hunted up the doctor with great trouble, but he seemed quite unconcerned. "Oh yes that's old so and so," he said, "He's always doing that".

5.06 MY THESIS

I had completed the experimental work on my Ph.D. thesis at Cambridge under Sir Denys Wilkinson, and started writing it on the boat to Australia – almost anything would distract me from this onerous chore, but it is difficult to get away from it on a liner. The first draft I typed myself. Four copies were needed, using carbon paper and strong fingers to produce them – no electric typewriters! Diagrams were a problem too – I used what was essentially blueprint paper. Having completed it, I needed to be examined – now here was a problem: my thesis director was in England, four weeks away by sea. However, there were some very ingenious people at Cambridge. They appointed Sir Ernest Titterton and Sir Mark Oliphant as examiners – the two individuals who were my bosses at Canberra. Not surprisingly, it proved advantageous to have them appointed, because it was in their interest to make sure I got my Ph.D. – after all, they had hired me. It was also in their interest to make sure it was a good Ph.D.!

Oliphant was a most remarkable person. During the WW2, it was his persistence that convinced the Americans to go ahead with the atomic bomb project. It is all detailed in the book "The making of the Atomic Bomb", by Richard Rhodes. Had it not been for this, the bomb would have been at least a year later in development. This was in spite of the fact that his personal war work was heading up the project which developed the magnetron oscillator for radar – what probably lead to winning the air war. So in a way, Oliphant had cut the time to win the war in both the east and the west.

While I was in Canberra, he was convinced to take the part of God in an amateur play production. Theatricals were an essential ingredient of life there – because there was little else to do, we had to amuse ourselves. With his shock of white hair, he made a perfect God – and it tickled him immensely to do it. Being director of the school, he was top dog, and it amused him greatly that Cambridge did not politely request him to examine me. They said bluntly, "You will examine" – this was because Oliphant was a graduate of Cambridge. He requested I shorten my thesis – so far as I am aware, this is the only time a thesis has been shortened. My examiners let me use a secretary to retype the thesis. Thank goodness! The thesis examination concludes with an oral examination, most of which is concerned with the topic of the thesis. However, traditionally, it concludes with a very simple question to which the student often gives the wrong answer. For example, a man rows to the middle of a small lake, and throws a rock overboard. The rock sinks to the bottom. What happens to the level of the lake? Most students would say (except that now this is a well-known problem) the level stays the same. Actually, it goes down, because in the boat the rock displaces its weight of water, and on the bottom, it displaces its volume, which is less. Anyway, Oliphant posed the question, "Einstein's general theory of relativity predicts that a ray of light passing close to the sun will be bent differently than under Newton's laws. What is the difference?" Brightly I answered, "A factor of two sir". This was the correct answer, and I was passed. What Oliphant didn't know was that, for the life of me, I could not recall which way the factor of two went!

5.07 RECREATION

Weekends we would go to the Murrumbidgee river in the summer, or skiing in the winter.

Normally, skiing, we stayed in cabins put up to accommodate the workers in the Snowy River project, which was just coming to completion, leaving them vacant. Cross country skiing was what we did mostly, since there were no tows or lifts in that area.Summer we would go swimming in the Cotter or Murrumbidgee rivers – there were quite a few snakes about, and we always wondered whether snakes could bite if you were in the water. So far as I know, we never found out. One of our party was the University House dietitian so we would take a picnic hamper from the kitchen – it was very informal, and go sliding down the rapids, wearing out our pants.

Then there was the famous Canberra wog. A few days after arriving, everyone got diarrhea – careful examination showed no bacteria in the water, and ultimately it was found to be a very fine sediment from the Cotter dam. Since the water was not filtered, you drank the (invisible) mud. It irritated the gut. Boiling the water helped by precipitating the mud.

5.08 THE WOOLSHED HOP

Once a year the sheep had to be sheared at the nearby stations (ranches). The shearers would come and stay for a few days, then going on to the next station – the most important person being the shearer's cook – the shearers must be kept healthy. After the shearing, the concrete woolshed floor was covered with the grease from the fleece. This was an ideal dance floor A notice would be put out that a hop would be held the following Saturday, and everybody and his uncle would attend. A good time would be had by all.

One time I tried shearing a sheep. You must hold it up while operating the shears. It requires strength, skill, and, in my case, a good deal of luck. One has to take great care not to nick the sheep with the electrical shears. Mechanical shears are a bit safer on the sheep.

One of my friends was doing research on cattle. He had to obtain semen from bulls, and I recall him telling me that after a while he had developed a peculiar relationship with this bull he milked of semen!

5.09 CARS

Cars in Australia were very interesting. Most people had a (General Motors) Holden. General Motors, after they had finished with the dies for pressing the body of cars manufactured in the USA exported them to Australia, where they were used to produce new cars, but of an earlier design. The first thing one did on buying it was to put a sack of sand in the trunk. This (to some extent) prevented the the car from overturning on a dirt road – it ensured the rear drive wheels gripped the road better. Nevertheless, three of my friends turned their cars over in one week – admittedly an unusual week. The top of the Holden was very solid, so no-one I knew got badly hurt from this. But driving along those wash board ridged roads was a misery – there was one speed (generally the one at which you wanted to drive) where the car vibrated terribly at resonance. It also skidded from side to side. At slower speeds, you had to hold your jaw tight closed, or your teeth juddered. Faster speeds, it skidded worse. Then, also the car filled up with the red dust rising from the road.

The first two years I had a motor bike – a BSA Bantam 125 cc – the dust was not as bothersome. Our student Don Gemmel had bought a 1929 Chevy. 1929 was a good year for wool, so all the farmers had gone out and bought Chevys. It drove very well on the rough roads, and sat so high, the dust ran out of it – so it was a good car, which I drove quite a bit. One day I hit a big bump, and the windscreen jumped out into my lap and shattered. I thought I would have a huge bill, but I went to the glazier, who cut a new one out of plate glass – it cost very little, and was easy to replace. Some of those old cars have advantages you don't consider.

Each car carried a canvas water bag on the front bumper. Water seeped out through the canvas, and evaporated in the dry Australian air as the car drove along, providing a delightful cool drink – better than air conditioning, which was not then available.

I later shared a Riley with Ted Irving. It had a pre-selector gearbox, a predecessor of automatic transmission. You selected your next gear to be used, then when you hit the clutch, it went into the new gear. Unfortunately, if you left it in gear when parked with the engine running, it had a tendency to creep forward, which it did one time, doing immense damage to a Hillman car parked just ahead of us. The Riley was unharmed

5.10 QUEANBEYAN

The alcohol laws in Canberra – the ACT (Australian Capital Territory) were pretty restrictive. However, those in the adjacent state of New South Wales were much more liberal. The village of Queanbeyan – a very old and interesting township, for drovers, station hands etc. etc. long before Canberra existed, was just over the border from the ACT. So after the bars closed in Canberra, the obvious thing was to go to Queanbeyan. Equally obvious, the police would wait on this single road to Queanbeyan when the bars closed there, and catch all the drunks on the way home.

5.11 LINDSAY TASSIE

As my wife Margaret's pregnancy advanced, we looked for a place other than University House for the baby, and finally found Peter Treacy, a colleague, was going on sabbatical, so we rented his house on Officer Street, near Black Mountain (which ultimately had a spectacular brush fire). We lived next door to the Tassies. Lindsay Tassie had just graduated Ph.D., and also just married. Lindsay was a night person, and being a theoretician, he would stay in bed till ten in the morning, and work late hours. However, this bothered his new wife, who quietly enquired of Margaret if this was normal behavior. She was used to a family where the man went to work at nine, and returned at five.

5.12 VISITORS

One interesting individual was professor James Meade – he was on sabbatical leave in Australia from Cambridge, a very pleasant elderly economist who dressed very formally by Australian standards, wearing a three piece suite. He was one of the founders of GATT, the General Agreement on Tariffs and Trade, and retailed some interesting stories about its early days. He was rather a shy retiring individual, a follower of Keynes. Meade, I discovered recently, was later awarded the Nobel prize for economics.

Another interesting guy who once was my next door neighbor at University House was a volcanologist. This is an individual who studies volcanoes. He was putting microphones up the sides of the volcanoes – but they suffered from severe static, because of the meteorological electrical action around the volcano. He hoped I might help him here, but I was of little use I am afraid. He would also fly around them in a small plane as they were erupting. Most interestingly, he had received a George medal from the king in WW2. He was placed on a small volcanic island, which showed signs of erupting. Every day he would climb to the rim of the volcano, peer in, and decide whether he thought it would erupt that day. Each day, there would be a bit more activity. Well, he ultimately came to the point where he felt they should not wait any longer, and the islanders were shipped away. Sure enough, a week or two later the volcano blew its top – for which he got the medal. Not, of course for the eruption, but for the prediction.

Because of the lack of very important people and high muck-a-mucks in Canberra, by default I got to meet some very interesting types – for example, I recall having dinner with Malcolm Muggeridge – he appeared to be a very irascible individual on English television, but was quite sociable and charming in the flesh – witty and entertaining, and not a bit aggressive. Then there was the British Prime Minister, Harold Macmillan. He pointed out that it was his family which published the first real "popular" scientific journal "Nature" – although, as he also mentioned – people who had never heard of it sent in articles thinking it must be a nudist magazine. Then came the famous Cambridge historian Arnold Toynbee a most erudite individual – quiet and unassuming – and of course, Sir Charles Darwin, who gave us a talk – he was on his way to New Zealand, in connection with the Rutherford centenary. The Darwin family is confusing. The men seem to be named George or Charles. They lived opposite me on Silver Street when I was in Cambridge. The Darwin and Keynes families intermarried, leading to progeny of exceeding brilliance.

One of the most interesting visitors was the Dean of King's college Cambridge. His name was Shepherd. I had known him very distantly in Cambridge. His voice was familiar as reading the lesson at the King's College carol service for many years. The tale went that he seemed very young as a don (on high table at Kings) until one day he said, "Tomorrow I will be old," – and, sure enough, the next day he arrived in a dark cloak, white-haired, hobbling along on a cane. He had a habit of patting me on the head and saying, "Bless you my boy," – which at Cambridge seemed perfectly acceptable. However, when he did this at Canberra, people looked rather askance.

Homi Bhabha the famous Indian scientist came by for a talk on cosmic rays at high altitude. He had one slide which, I thought, showed the difference in attitude between east and west. Nuclear emulsions – photographic plates sensitive to nuclear particles – were flown to great altitude to avoid the atmosphere. In America, they had developed gigantic balloons, and the technique for filling and releasing them. However, on this slide, Bhabha had a large field covered with Indians sitting cross legged about two feet apart. There were literally hundreds of them. Each Indian was holding what looked like a hydrogen filled party balloon. On command. the first row would release their balloons, then the second row, and so on – and the balloons were all fastened together, so that as the last one was released, the package of plates would be lifted majestically into the sky.

Harold Bailey, the famous Sanskrit scholar came through. He had been a next door neighbor at my college at Cambridge where he played the cello in quartets. Starting late in life, he had decided it was more productive to learn the cello than the violin. He was an expert on languages, and was reputed to remark, "When you know thirty languages, learning another is very easy – they all have the same pattern". He was a very mild mannered man, but I have a delightful picture of him dressed as a brigand. It turned out he gave a talk in the Balkans, and the audience was so impressed that he spoke their rather obscure language, they presented him with this local dress, sword, bandolier and all.

Curious things happened. One time, a salute was fired at Duntroon Military Academy, (the equivalent of West Point, or Sandhurst) for a visiting dignitary. The wadding for the cannon set the grass on fire, and everyone had to go and put out the brush fire that resulted. Mt. Stromlo almost burned down as a result of such a fire – everyone was impressed to help out. More recently, the observatory did burn down because everyone was on holiday when the bush fire occurred.

5.13 MARRIAGE

There was no television in Canberra, and not much radio. The announcer was a student with whom I frequently had lunch. So we were thrown on our own resources – play reading and folk dancing have since then been a large part of my repertoire.

Furthermore, the residential dorm was coeducational, at a time when such a thing was unheard of in either England or the USA. Suffice it to say this led to an interesting environment – and quite a few marriages, one of which was mine. I well recall Mick Borrie, my wife Margaret's boss and advisor – who gave the bride away – saying, "When two Australians marry in Australia that's marriage – but when an Englishman marries a Czech in Australia – that's definitely intermarriage". As you might expect, he was professor of demography, and, I believe, the first such professor in the world. The press, never having heard of demography, frequently referred to him as "professor of democracy". We were married in St. John's church by the coadjutor bishop, Bishop Arthur, and the whole thing was recorded on tape by my technician and transferred to disc. The future prime minister of Australia (Bob Hawke), and the American ambassador attended the wedding – but that was not such a big deal – Hawke was a lowly post doc, and there was very little other entertainment in Canberra. As a result we were thrown very much on one another's company – and since the town had only about 30,000 inhabitants, this meant that we entertained the legislature and the embassy staffs – because that was what the town mostly consisted of. There was one federal treasurer who loved parties – so when the students had a party around the piano in the basement, you would find him there. In earlier days it was even wilder, so I was told.

There was the famous tale of the steamroller race. Two drunken politicians came across two steamrollers used to level the roadway. They set them going – maximum speed about four miles per hour – and had a race. I well recall while I was in Canberra one intoxicated politician getting up in the House and rambling on. Another rose and said "You're drunk." The first replied "I am not (hic)". "Yes you are," was the rejoinder, and this interchange went on for quite a while. This was when they decided to broadcast the speeches in the house. The result was that when the people from NSW were listening, their legislators spoke. As this time zone passed across the country, the members from each region would speak.

We had very good meals – so many of the politicians would come around for the grub – Australians are very forthright – and mostly honest. But beware drinking with them. The Australian beer is excellent – but also strong. Shortly after I got there, one of my colleagues had the misfortune to die of misdiagnosed diabetes. We went to the local pub for a wake – and I had one beer for every two the others had – and even then I was completely crocked by the end of the evening – but it was a very good wake. The deceased would have enjoyed it.

We had some very good fancy dress parties – again, with few other distractions, we could make up the most exotic scenes. Oliphant's nieces used to come regularly, and I recall dating one – after I got my PhD.

Many important people gave talks in Canberra. Since it was the capital of a country the size of the USA, they felt assured of a big audience – however Canberra was small at the time – but this meant you were sure of a good seat whenever you went. I recall sitting next to a very tall, distinguished and elegant white haired statesman (Sir Robert Garran) at a talk given by the American supreme court justice William O. Douglas, one of the most controversial Americans, and quite a character. Anyway, the talk proved rather dull, so the statesman simply turned off his hearing aid and went to sleep. Luckily, he didn't snore.

There were several students on Fulbright scholarships, and they would receive invitations provided by the ambassador from the USA. Since this was the source of their funding, they could scarcely refuse. However, it turned out that the ambassador liked to play charades, and because the embassy had few people, he recruited these students – who would come back later grousing "what a waste of time". Of course, ambassadors are a political appointment in the USA, and Australia was considered safe – not a likely source of problems – so we did not get the most impressive ambassadors. One had the misfortune to die and be buried in Canberra.

5.14 OFFICER STREET

After getting married, we lived in an apartment in University House, but Peter Treacy then rented us his house in the suburbs – a prefab – when he took a sabbatical in Oxford. I was driving a Ford Prefect, made in England, not a good car for the Australian climate, because it has a mechanical fuel pump, which, heated by the engine, boiled in the summer, creating a vapor lock. One had to get out and put a wet rag on the pump which cooled it enough to condense the blockage. We spent some time down at the coast near Bateman's Bay in a cottage belonging to the head of the anthropology department . The first night there was a thump thump-thump on our veranda – mysterious – in the morning we went out and found a bunch of kangaroos. One had jumped the barbed wire fence, hopped along the veranda, then jumped the other fence to get out.

Our honeymoon was actually spent at a place called The Entrance, north of Sydney, which had a nice beach and was not too far from the Blue Mountains, which we visited. It was pretty chilly. The view across the rift valley was spectacular.

The Officer Street house depended on a wood fire for heat – very inexpensive because the wood was provided by the foresters of the national forest. We were afraid a spark might set the house on fire, so to preclude this, I spread explosive fireworks on the floor, in hopes they would waken us before the fire did much damage. Since the baby slept in the living room, where it was warm, this was important.

Margaret had to be hospitalized for a ovarian operation just after Christopher was born, so I was in charge of the baby. The poop provided by a baby fed on formula is appalling, and I had to rinse out the nappies, and boil them in a massive copper pot or kettle. They had then to be hung out to dry. We said Chris was a second hand baby, because all his baby clothes were second hand, since I couldn't afford new. Still. it was basically a happy time. The house had a metal roof, which Peter had arranged to have running water spray on to keep it cool. Most old Australian houses had corrugated iron roofs, painted red to inhibit rust. Nothing is more soothing than the sound of rain on a corrugated iron roof.

5.15 THE SYNCHROTRON

The accelerator arrived in Australia in parts a couple of months following me. It traveled in crates, and seemed to be in pretty, good condition. Neither I nor my colleague, John Carver had ever run such a machine, so we set about erecting it with some trepidation. We had drilled a hole in the basement wall as a beam dump, and hoped the earth around it would be enough shielding. Most of the lifting was done by an Irishman – inevitably called Paddy. He had a stronger accent than any Irishman I had ever come across previously, which leads me to generalize that people emigrating tend to emphasize their original characteristics – the Irish are more Irish in Australia than Ireland, and I confess I myself – though not deliberately, may be subject to this problem. I am more English abroad. People in America have accused me of returning to England to refresh my accent. On the other hand, I went into a shop on the Town Hall Square in my home town to buy a gramophone (phonograph) record. Returning the next day to pick it up, the girl behind the counter said, "That American is here again." I felt quite deflated. I recall thinking Alistaire Cooke was American – till I heard him after I got to the USA.

We put the machine together and hooked it up. Since it ran at 11,000 Volts AC, we were cautious. Luckily, England and Australia have the same electric mains frequency. We ran up the volts by adjusting an enormous shunt choke. As the voltage went up, we heard "bang, bang, bang". What was it? It turned out to be the magnet pole pieces hitting one another – we had to put the vacuum chamber or "doughnut" in to hold the poles apart. After that all went well. Since the synchrotron worked by resonating its inductance with a bank of gigantic capacitors, it took vast current, but the current was 90° out of phase with the voltage. The net result was we consumed little energy, but lots of current. Since the line losses depend on the current, we thought the power company would get very mad with us. Strangely enough, this was not the case. It turns out that most devices, motors etc., are inductive, and in order to bring the current and voltage back into phase the power company must install large capacitors. Although we were out of phase, it was in the opposite sense, so the power company was quite happy with us.

The acceleration of the electrons took place in a glass "doughnut", and we needed a spare. We got it from England, but in order to speed delivery, we had it shipped in the diplomatic bag, via the embassy (or rather, Office of the High Commissioner). What we had not realized was that the bag contained a large lead weight. This was so that if the plane crashed and broke apart, the pouch would sink, instead of floating and being found by the enemy. The lead weight hit the doughnut, so all we got were a few broken pieces of glass.

The capacitor bank would occasionally spark over, with the most immense bang. We would leap up, but irrespective the machine continued to run. One time Paddy was down there when it went off. We thought now we would find out where it sparked. Paddy however said it was like lightning, running round the room. On thinking about it, what must have happened was that the bright spark etched a spot on his retina, which as he jumped, seem to run all over the place.

5.16 RADIOACTIVE COCKROACHES

We had radioactive sources around the lab, which were stored in a locked cupboard. Since a fair amount of radiochemistry was performed, periodic checks of the lab were made. However, even when no chemistry was being done, radioactive material was found, in small quantities, distributed in the most bizarre places – under the sink etc. This was a severe puzzlement – was there a leak, a spill or something? Finally the whole thing was cleared up when someone coming in late at night, and turning on the light, noticed cockroaches scurrying under the door of the radioactive cupboard. They were eating the radioactive material, then emerging in the dark to distribute it all over the place. Insecticide, and a closer fitting door cured this.

5.17 DON GEMMEL AND THE COCKCROFT WALTON ACCELERATORS.

One of the projects in progress when I arrived involved the two Cockcroft Walton accelerators. These were virtually identical to the ones I had used at

Cambridge. However, the larger accelerator was supposed to work at well over a million volts, which it never did. The reason was that when it was bought, the fact that Canberra's altitude was in excess of two thousand feet had been forgotten. Since the accelerator had a high voltage terminal which was air insulated, the lower air pressure meant that the terminal sparked over to the wall at a much lower voltage. I used the lower voltage accelerator to generate gamma rays with a new graduate student named Don Gemmel. He later became head of Argonne National Labs Physics Department, in Chicago. We ran immensely long hours on this machine, and to occupy the time late at night, I brought in a trumpet to practice on, and Don brought in the chanter from his bagpipes. Later, in Chicago, he used to judge bagpipe competitions, until he quit, because it was driving him nuts. The competitor would come in and march up and down playing the bagpipes while he judged. Can you imagine hearing the same bagpipe melody played badly by innumerable competitors? It sounds like a form of torture to me. Anyway, we were playing at about two in the morning when the guard marched in, accompanied by his Alsatian – a nice dog. It turned out he was a trumpeter until he "lost his lip," something I had never heard of before. He gave us a performance on the trumpet, and we had quite a nice time. Forty years later, I had dinner with Don and his wife at his home in Chicago when I was running on the Argonne pulsed neutron source with Vicki Homer and her son – only time I researched with a family.

5.18 THE FIRST SATELLITE

It was announced on the news one day that the Russians had shot up the first satellite – Sputnik. We had long discussions with the astronomers from Mt Stromlo as to whether this thing would be visible. They had varied opinions, but none of them really seemed to know. Later on, my next door neighbor, Lindsay Tassie and I went out to see if the satellite would be visible at dusk. Sure enough, we saw it, going brighter and dimmer as it passed over. Lindsay had a cheap Brownie box camera, which had a time exposure stop – so he opened the shutter and put it on a stump. Amazingly, when developed, you could see the track of the satellite. And Lindsay was a theoretician! In connection with our radio astronomy work, we had a very talented astronomer, Mort Roberts, who noted that the Russians had given out the frequency of the Sputnik. So he put together a Yag antenna with wire and a piece of spare timber. As the satellite went over, you could point the antenna in the right direction and hear it beep. Marvelous. Roberts had just arrived in Canberra, and brought with him the best beer – Coopers of Adelaide. Ostensibly, the brewer had died of drinking too much of his own beer – a great recommendation!

5.19 COSMIC RAY NEUTRON EXPERIMENTS IN CANBERRA

I had left England glad I would never again have to perform another experiment with potassium permanganate, a miserable substance which stained everything a deep mud color. By some mysterious means however, I was convinced by the prof. that a vast uncharted sea of cosmic ray neutron experiments lay untapped at my feet, or rather, to be more specific, thousands of feet in the air. Unfortunately, I got inveigled into doing experiments on cosmic ray neutrons using the same Szilard Chalmers reaction I had employed in England. It was a bad day when I accepted – but nevertheless interesting, in the same sense as the old Chinese curse "may you live in interesting times". I had a setup employing a beer keg of concentrated sodium permanganate solution. This stuff was such a strong oxidizer, it set on fire anything it fell on. We succeeded in getting one of the Royal Australian Air Force's DC3s – (C45 in America, Dakota in England). I went up with the pilot to see if the plane was suitable, and he proceeded to put it through aerial maneuvers which were little short of loop the loop. As I recovered from my queasy stomach I asked him why he had done this. His reply was, "Didn't you want this?". It turned out the previous scientist to employ the plane had been seeding rain clouds with silver iodide crystals. For this, you have to be able to get the plane here and there wherever the clouds are, which involves sharp turns and maneuvers and so forth. I quickly disillusioned him, and told him that we wanted the plane to fly as stably and evenly as possible.

After that we went up a few times with the apparatus. As we were loading it, the keg of permanganate would inevitably spill a few drops of the vile solution on the wooden floor of the aircraft, where it would burst into flames which would be rapidly stamped out. My technician and I would take turns distracting the pilot while this was going on, so that he would not be upset. Ultimately, we had to take the plane up to 30,000 ft. Now, no mountain in Australia is above about 7,000 ft, so there was no need for oxygen, although, these being military planes, it was available. Anyway, the crew charged the oxygen cylinders, and we started up. The superchargers kicked in at altitude, but the rate of climb was relatively slow above about 20,000 ft – these planes were from the thirties, and of course, not pressurized. My technician would take a deep breath, then dash back to check the equipment – oxygen was only available in the cockpit. Well, we got above 28,000 ft, which is pretty good in a DC3, when I noticed my mask was giving oxygen even when I was not breathing. Looking at the pressure dial on the oxygen cylinder, we noticed it rapidly decreasing. As it reached zero, the pilot said "I think we had better go down now". We descended much more rapidly than we went up!

The filtering of the solution occurred both in the air and on the ground. One time, luckily on the ground, in the small room assigned to us at the airport, the hose broke, pumping a fountain of permanganate all around the room. This happened when my assistant was in charge. I was at a party at the time, when he suddenly appeared. The permanganate, although relatively harmless, stains one a dark brown, so he was piebald, and looked as though he had some terrible disease, disturbing the guests at the party.

This was not the end of the permanganate problems. In order to study the effects of water on the diffusion of cosmic ray neutrons, I had made a shallow sealed container full of the solution. A suitable large water tank was the enormous cistern in the top of the tower of the physics building. No-one ever went up there, so I was happily doing experiments. However, Peter Treacy had a pressure vessel he wanted to test. Why not put it in this tank, and look for bubbles? So up he went, not knowing about my tank of permanganate, which he proceeded to disturb, and spill. It turned out there was no way to empty this big storage tank, so the whole water supply for the department ran purple for at least a week. The stuff is quite harmless – in fact campers use it to sterilize drinking water – but it looks lethal, and the women on the staff particularly were quite put off. Years later, Bill Turchinetz gave a talk at USC, and reminded me of this event – it must have made a deep impression.

When I finally left Canberra, Mark Oliphant, the head of the department, presented me with a toy Koala bear, died a deep permanganate color.

A bizarre sense of humor was one of the requirements at ANU. I recall being in a room doing research with Don Gemmel one time. Looking out the window we found we could see a friend of ours in the medical school at his desk – but he could not easily see us. Don picked up the phone and dialed his number. Just as he was about to pick up the phone, Don put his down. This was repeated about three times, by which time our friend had his hand directly above the phone ready to take it when it rang. Then he gave up, so of course, we let it ring a long time before he was forced to pick it up again. We never told him about it, and I have often wondered what explanation he had of this phenomenon. It is of such material flying saucers and little green men are constructed.

5.20 THE MEN'S ROOM

One of the interesting facts of life consists of running into famous men as they enter or leave the men's room. The reason for this is not difficult to see. No matter how famous (or infamous) you are, sooner or later you have to visit the men's room – (unless you are a woman!) whether it is in the airport terminal, restaurant or theater. (George V said "never pass up a men's room if you see one". Evidently, people never thought he might need to go!). Generally one is in a hurry and bump into others. In this way I bumped into the Duke of Edinburgh when I was at the ANU in Canberra, and David Brinkley in Atlanta airport. Both were much taller than I had thought. In spite of this interesting observation, it is probably not a good idea to keep going in and out of men's rooms on the off chance of running into someone famous. It might be thought one has a peculiar urinary disease.

5.21 THE BIG MACHINE

Oliphant had the bright idea of building the world's largest accelerator in Canberra. This was to be run by a homopolar generator, a device based on "Faraday's disk". The concept was ingenious, but Oliphant had not reckoned with the labor problem in Australia. Basically, four twenty ton twenty foot in diameter steel disks were to be placed between the poles of an immense electromagnet. By feeding current into the periphery, they could be spun up until they did not quite burst. Then they were short circuited through the coils of the gigantic air cored magnet which formed the actual accelerator. A vast magnetic field was produced, which ostensibly would be used to accelerate protons to GeV energies. It never did work out – the concept was preempted by the "strong focusing" principle devised by Courant and others in America. However, the gigantic magnet proved later to have several useful purposes connected with plasma research. Prototypes of the generator were produced using NaK as a conductor. It is a dangerous mixture of sodium and potassium which is liquid at room temperature. This was squirted at the edge of the discs, to produce an electrical contact. If water, even the smallest quantity, got in the NaK, it burst into flames from the hydrogen produced. Anyway, this did happen, causing much trouble. One of the discs now stands as a monument outside the ANU physics building. To avoid the NaK, special carbon brushes were devised to handle the high current into and out of the disks, and were quite successful.

5.22 VISITING SYDNEY

During my sojourn in Canberra, I had to stay in Sydney at various times. One time, I was staying at Fernanda de Carvalho's mother's house on Double Bay. A large American aircraft carrier was anchored in the bay, and I was awakened in the morning by the Stars and Stripes coming to me across the water.

After breakfast each day I would stroll down to the dock, and be picked up by the ferry, which transported me at a leisurely pace to Circular Quay, whilst I read the paper, or admired the view. I could walk from here to my place of business. I wonder if it is still possible to do this? It was a truly civilized way to get to work.

6 AUSTRALIA TO BLIGHTY

6.01 THE SOUTHERN CROSS

Our return to England, in the fall of '58, was via the Panama Canal on the Shaw Saville liner the Southern Cross. We thus circumnavigated the globe going East. For future reference, I recommend going west. The reason is, you have to get up earlier each day going east, with the benefit that as you cross the dateline, you have an extra day – for example, you might have two Wednesdays, or two Thursdays together. So, I am one day older than other people with the same birthdate. The extra day does not compensate for having to get up earlier. Going the other way, you can stay in bed the extra half hour. This strikes me as being much more civilized.

The Southern Cross was a most unusual ship. It was the first liner which had the engines in the stern. This was a big advantage, because it meant you could have a large room in the center of the ship, without funnels, etc. coming up through the middle. This room could be used for movies and dancing. Dancing in such a large room proved a delight. As with the boat I was on going to Australia, we would dance downhill when the boat rocked in one direction, and downhill the other way as it rocked back. However, the effect was not as large as on the old Orion, because the new liner had stabilizer fins. The boat would start to rock, and then suddenly stop as these fins tilted, and took effect. Then back the other way. I am not sure whether this was better or worse than having no stabilizers, which gave a more gentle rocking motion. The other advantage of this ship was that it had only one class. This meant two thirds of the passengers were not restricted to about a third or less of the ship – you could go anywhere – including the nice pool. It was not a large boat, but brand new.

6.02 INTERESTING PASSENGERS

There were two interesting people on the boat. One was a returning administrator of a British colony in Africa. He had been at my college in Cambridge at the same time I was. His colony was adjacent to a Belgian colony. This proved advantageous, because the Belgians exchanged wine for British products ("I wonder what the vintners buy one half so precious as the stuff they sell?"). Unfortunately, he had contracted chicken pox, which somewhat ruined his trip. The other individual was a Mrs. McWhinney. She sat at our table. Her younger son had been a close friend of mine at high school. He met a bizarre and tragic end. While playing field hockey one day, he was struck on the head by the ball. He passed out, but appeared to recover. A week later, he was in a play, and dropped dead on stage. It seems the impact of the ball had severed blood vessels in the brain, which later gave him something like a stroke. At first the audience thought his demise was part of the plot.

6.03 WELLINGTON

Setting off from Australia, our first port of call was Wellington, New Zealand. It took a good five days from Sydney to Wellington – it is much farther than most people realize. You pass through the Cook straits between the North and the South Islands. In the old days, you were guided by "Pelorus Jack," a famous white dolphin – but we only saw the normal kind. The parliament house was built of wood, because of the prevalence of earthquakes. It was felt the wooden buildings would be less subject to damage, and less likely to hurt people if they collapsed. We took an inclined railway up a hill to a vantage point over Wellington, and I recall a tour including the black sand beaches.

6.04 FIJII

Next came Suva, Fijii (our ports of call were all British possessions, or pick up points for oil). I recall trying out their famous native drink, kava. It has a curious astringent flavor, leaving your mouth feeling clean and dry. After drinking it you walk around with your lips pursed for quite a while. The Fijians had a competition to see who could open a coconut most quickly, using a machete. First they chopped off the outside fuzz, then chopped a hole in the woody part with this ferocious looking machete. The milk tastes good. We took a tour of Viti Levu, the volcanic island on which we landed. The Fjians are very easygoing, but the British imported hardworking Indians, and the competition between the two has proved a problem.

6.05 TAHITI

Next stop – Tahiti. There is to my mind, no more beautiful view than that of Tahiti, seen from the sea as you approach at dawn. The main island is surrounded by a coral reef and there is a narrow channel to be navigated, with great care, to enter the lagoon. I had always admired Gauguin's paintings of Tahiti, and had thought it impossible that the deep purples and magentas he used could exist in reality. Yet in fact, the paintings could not compare with the reality. The craggy peaks of the volcanoes in the distance provided a stark background to the lush palm tree laden foreshore. It was awe-inspiring, and the colors changed continuously as the sun rose, until they became much less complex, and as we docked, it seemed like a normal port. Docking, at that time in the port of Papeete was not easy, and our ship had tubes transverse through the hull with propellers within to pull the ship towards the dock, off which it lay stationary a few feet away. Had it tried to slide in parallel to the dock directly instead, the physical Bernoulli effect would have created suction to make it collide with the wooden wall disastrously.

You must remember that at this time, these islands were practically inaccessible by air, but since transportation by water was then much more common, there were many liners, not cruise ships, but plying between Sydney and London, or the many parts of the British Empire as it then was. Tahiti, however, is French, and we arrived there at the end of the Bastille week – so there was still a lot of celebrating going on. The Tahitian girls are exceedingly beautiful when young, but blossom out to a much larger size as time goes on. Luckily the Tahitian men like it that way.

We were told that people rarely go out much before noon. The reason for this is that there are many coconut palms, and the coconuts drop off the trees in the early morning, so it pays to stay in till late in the morning. This seems like a good excuse. The groves did not have "Trespassers will be prosecuted" on them – but just "Taboo".

My first requirement, on landing at Papeete, was to get condensed milk for the baby. It had not struck me until that time that people speak French here. The shopkeepers for the most part were Chinese. I entered this store and thought "What is the French for condensed milk?" I tried "Lait condensé". The shopkeeper looked puzzled. Just then I saw cans of Carnation milk on the shelf. I pointed to them and said "Ca." "Oh," said the shopkeeper, "Le Carnation!"

We were shown around the island by a young lady purporting to be a Tahitian princess. There must be several of these around, and wearing a flower coronet, she certainly looked the part. She was both beautiful and interesting. She said that, in addition to being an island paradise, which Tahiti is, it was also very boring. If you grew up on the island of course, this was not the case, but people would come from Europe, find it deadly dull, get fed up after a year, and leave. Since it was two or three weeks by sea to anywhere from there at that time, this made good sense. Now, with a large airport and rapid transport too and fro, this is probably not the case. One of the celebrations involved fire walking. The participants walked on a bed of hot coals. Being of a timorous disposition, I did not take part. Jearl Walker, a physicist I know used to do this. The secret is that you are scared witless at the thought of walking on the hot coals, and the sweat protects your feet so they do not burn. Ultimately he became blasé, his feet did not sweat, and they carted him off to the hospital.

We went right around the island, which has a volcano in the middle, and saw a small waterfall where their religious ceremonies were performed. These were rather strange to western views. In the past, one of the problems was overpopulation – such a small island could rapidly become overpopulated, leading to poor living conditions. Hence, babies were left out to die, but as part of a religious ceremony, which, I believe, took away some of the trauma involved.

Our baby Christopher had become a toddler on the boat, learning to walk. However, his first attempts on shore were interesting. He was used to compensating for the ship's motion on board by wobbling from side to side, which lead to a straight path. On shore however, he continued this compensation, with the result he weaved a snake-like path.

6.06 PANAMA

From Tahiti followed a large stretch of ocean to Panama. We stopped off at Panama City which had been razed by the pirate Captain Henry Morgan. The ruins of the old city were most interesting. I remember standing in the doorway of the destroyed cathedral. The city itself was very picturesque, with balconies a bit like New Orleans. The canal too proved delightful – you could see monkeys in the jungle. We waved our glasses at a USA navy ship. Unlike the British at that time, the crew were not allowed alcohol. The lock houses had roofs which folded back. These control buildings lie between the locks. American aircraft carriers in the old days were built to pass through the canal. However, lack of forethought meant the designers had not allowed for the large overhang of the carrier decks taking off the roofs of the nearby lock houses. After some thought, these were altered to fold back, so the decks just missed the roof. The large hinges were clearly visible. We passed under the bridge of the intercontinental highway.

6.07 Curaçao

Next stop – Curaçao. This typical Dutch town, with its gabled roofs, drawbridges and canals, just like Amsterdam, was a delight. The boat gassed (or oiled) up there, and we took a tour. The taxi driver told us more than we wanted to know about the island. Being part of the Caribbean, we were serenaded by a troubadour with a calypso. He made up a suitable calypso about Margaret, Christopher and myself, which proved a humorous delight.

6.08 TRINIDAD

Trinidad, whose capital is Port of Spain, was next. Its connection with the Caribbean, with romantic pirates, etc., was fascinating. We toured the capital and its market. We also took a trip to a beautiful tropical beach on the other side of the island. Christopher had never seen a beach before, and it was lovely to play in the sand and sea. Just watch out for the coconut palms!
After this we went straight to England, although we passed Tobago closely, and went through the Sargasso sea. This was mostly sparse weed covering the whole ocean, but not the dense islands we had been led to believe. We docked in Liverpool, and were met by my parents.

6.09 ENGLAND AFTER BEING AWAY FOUR YEARS.

It was nice to be back in England, but there were some changes. Everything had gotten smaller – the roads, the houses – of course this was the result of being abroad in Australia, where everything is bigger. The health scheme meant when we saw Dr. Wright my old doctor, he didn't charge anything. There is no doubt the public health scheme is a good thing – the English complain about it bitterly, but ask anyone would they go back to the old "panel" system – not under any circumstances.

We left from Liverpool on the Brittanic – the last Cunard motor vessel, which was driven by gigantic Diesel engines. It was kind of sad to see the statues of the mythical Liver birds on the Liver insurance building vanish over the horizon – the last sight many immigrants have of "Blighty" – the old country.

7 CROSSING THE ATLANTIC AND EARLY LIFE IN AMERICA

7.01 CROSSING THE POND

I had been interested in the University of South Carolina because the department chairman there was Tony French, whom I knew well from Cambridge days. He had been attracted to South Carolina as a teaching position, since he felt he might be more interested in this than research. Shortly after Tony arrived here, the department chairman, Fred Terry Rogers, died. Having tried to mow his lawn on a hot summer day, he suffered a heart attack (he was a northerner, and did not realize the dangers of physical activity in the hot, southern climate). Tony accepted the position of chair, and offered me a position when I wrote to him.

The trip from England to America proved rough. On the way over, the ship bypassed two hurricanes, with the result it met huge seas. Although we were fine, still having our sea legs from the Australian trip, everyone else was violently sick. We wanted to eat in the cabin, because our two year old had a habit of throwing food all over the place, but the steward would not hear of this. However, after a day or two of cleaning up in the dining room (which was deserted, everyone else being sick in their cabins) he let us do this. It turned out, our table mates were a couple who had quite a good sense of humor. The guy came in one day with a bowler hat – from a bygone era even then.

7.02 HURRICANES

The hurricanes – or rather their remnants – delayed us – we had to sail as far south as North Carolina to avoid the worst swells. This delayed us two days and we arrived in New York on labor day – and, if there is one day on which not to arrive in America, that is it. Strangely enough, the customs official who saw us through with all our bags, was extremely helpful, as was the cab driver – and it has always affected my attitude to these people ever since. He asked my weight "ten stone five" I said. Didn't phase him a bit. He wrote down "145 lbs". We went to the Wellington – a fairly cheap hotel in downtown Manhatten, and spent the night there. The cab driver rode us around a bit to see the sights. The next day, we took the Palmetto, the train to Carolina. There were then two crack trains, the Star and the Meteor, with air conditioning, and so forth, and then two from a bygone era, one of which was the Palmetto. With no air conditioning, at the end of summer, with, to us, incredibly high temperatures and humidity, it was a miserable trip. The carriage smelled too. Anyway, we were met at the station in Columbia by Ernst Breitenberger, a friend from Cambridge days, and Fred Giles, another faculty member.

7.03 BREITENBERGER

Ernst was a most interesting guy. When he was a graduate student at Cambridge, I was present when he and our technician talked about their days in the North African Campaign of WW2. They were on opposing sides, England and Germany. Ernst would say, "But we thought you (the British) were here", and the technician would say, "No no, we were here, but we thought you were there," in an entirely different place. In fact, to listen to them, it seemed surprising they were not still lost in the wilds of North Africa. Ernst had been captured, and spent the duration of the war in the United States. Our two families formed quite a united nations back in Carolina. He was from Austria graduating from Graz. After his second Ph.D. from Cambridge, he was "Herr Doktor Doktor" in Germany. His wife was French and at least one of their children was born when they were in Singapore. I was English, my wife was born in Czechoslovakia, and our first child was born in Australia. When I acquired ulcers, one in the stomach, and the other in the duodenum, whilst I was in California, he commiserated, because he had had a large part of his stomach removed for his ulcer. He said it had one compensation – maybe. If he took a drink it was absorbed immediately – no waiting around in his stomach, since it didn't exist – so he got an immediate high from the alcohol! Later Ernst left for Ohio.

7.04 FRED GILES

Fred Giles, who also met my train, was a very tall very American American from the mid-west. He was without doubt one of the nicest people I have ever known. He took us for our first A&W root beer. It seems perhaps odd that this would strike us as such an important thing, but it did. The cool frosty glass, the unique flavor, being served in the car instead of inside the building – the root beer place is long gone, I am afraid. There is nothing quite like getting used to the habits of a new country. Such an experience has always filled me with awe, and with a soaring sense of joy quite indescribable to a local inhabitant.

Fred was a remarkable individual. It always amazed me his schizophrenic ability to teach the fundamentals of physics, and yet believe in the fundamentalist Bible concepts. During the week he taught the "Big Bang" theory of creation, but Sunday it was 4004 BC, or whatever.

He had a pilot's license, and we would fly together periodically. Once we went down to a meeting of the Southeastern section of the American Physical society at Tallahassee. On the way back, the weather set in. We were flying visual rules, so we could not go into the clouds. However, the clouds got lower and lower until we were barely flying at tree top height. It was getting quite eerie.

Anyway, luckily we were approaching Valdosta Georgia, which had a small airport, and we landed there. In those days, each township had a water tower, with its name painted on it, and reading this, you knew where you were and did not get lost. Other pilots were stuck on top of the clouds, and it just so happened the clouds opened above Valdosta for about fifteen minutes, during which time five or six planes descended and landed. We spent the night in Valdosta, whose sole entertainment appeared to be a roller skating rink.

Fred's dedicated religious views were a help to him later, for he got cancer (probably while on sabbatical leave in Iraq), and I watched him die, visiting him in the hospital. It completely changed my life, because for many years I had been having a miserable time with my marriage, sticking it out under the belief that life was like that. However, watching Fred made me realize that I too could be dead tomorrow – we only live once, (I like to point out that some people don't even do that!) so there is no point in suffering. It ultimately led to the breakup of my marriage. I had heard that after separating, one got very depressed. In fact, just the opposite was the case, and I have lived a much happier life ever since. I am sure my ex-wife would have disagreed, but I find it difficult to assign guilt in these cases. Having experienced it, to say some sultry woman lured me away is quite false – there were women, but that was not the cause of the breakup. Since then, I have been very suspicious of anything I read on this subject, because people talk with such conviction about something I am convinced they know nothing about.

7.05 FOOTBALL FOUL

I had bought a pair of binoculars on the boat, and, wanting to try them out, wandered down to a large field with footballers playing, adjacent to the house we had been lent by my colleague Ed Lerner. I had been playing with the binoculars, adjusting the lenses, etc., when a large and burly man approached and asked me what I was doing? I explained to him in my English accent, and then he told me that this was their practice field and they were trying out plays for an important football game, and having people observing them meant their plays were being stolen. I apologized and told him I knew nothing about American football – apparently I was lucky he had not attacked me at first! My accent probably helped.

7.06 GULLAH

Gullah – or Geechie – are the two dialects spoken by people of African descent living on the barrier islands off the Carolinas and Georgia. A week or two after arriving in Columbia in 1958, we went down to Charleston. At that time there was only a two lane road, and I was puzzled to see a large billboard as we drove into Charleston, just a few miles out, which proclaimed "The KuKlux Klan welcomes you to Charleston". Knowing very little about either the KKK or Charleston, this was a mystery to me. It still seems ambivalent – but maybe not!

In Charleston we asked a black guy at a corner the way to the battery. He was very helpful, and said, "Ahwa ricko, yall," and so on. I said "I didn't quite follow, could you repeat it?" So he went on, "Ahwa ricko, yall," I had never heard of Gullah – but that's what it was and totally incomprehensible to me. So I thanked him profusely, and went on.

7.07 HENRY'S

Charleston believed it was above the SC law – certainly as far as alcohol was concerned. I went into an establishment called "Henry's" off the market square. It is still there, but is a different place entirely now. It was at a time when South Carolina was dry as a bone – no liquor could be sold in a bar. However, Henry's had an open bar, and I was astounded to find you could buy a Scotch, poured from the bottle, as you could in states without such restrictive laws. In Charleston, if they thought the law was wrong, it was just ignored. Ultimately, the state realized you can't stop people drinking, and we have gone through quite a few changes. The first was the "brown bag" law, which was a disaster. The law allowed you to buy a bottle of liquor in the store, put it in a brown bag to render it inconspicuous, and carry it into the restaurant for consumption. Since the smallest bottle you could buy was half a pint, it was natural to finish up the bottle before you left, with the result you would see somewhat inebriated diners totter out of the restaurant, with the distinct possibility they would have an accident later. More recently, the "mini bottle" regime has occurred. You buy liquor in a restaurant by the mini bottle which contains approximately 5 cc. Pretty soon, this too will vanish as we enter the era of civilized drinking under yet another law.

7.08 TONY FRENCH

As I mentioned, I went to South Carolina because the department head, Tony (A.P) French was an old friend in Cambridge, England. He had worked on the atom bomb project at Los Alamos, and I had always wondered how an English graduate could have been spirited to Los Alamos like that during the war. It turned out it was due to Bretscher, a nuclear chemist, who needed an assistant when he was drafted to Los Alamos. Nuclear physics being a brand new subject, you had to take trained people from wherever you could get them, which is why they probably inadvertently recruited spies. Tony met his wife Naomi there. She was Richard Feynman's assistant, as a mathematician. Feynman was in charge of certain bomb calculations. Being so isolated there, many babies were born, and Feynman had a notice board on which he stuck the cigars given him for a new birth. He would smoke the cigar till he felt sick, then pin it to the board.

I later learned from Naomi, that as Feynman's assistant she was in charge of the IBM digital computer – a very early one. It consisted of two parts, one which added, and the other which multiplied. They would do all the multiplications, then take the Hollerith cards over to do all the additions, She was bothered by a strange character – "Johnnie" working on the bomb project, and asked her boss what to do. "Explain it to him," he said. "Maybe he'll go away". She did, and he did. It was John Von Neuman, later a famous figure in computing.

My thought was to spend a year at the University of South Carolina and then go elsewhere. However, I enjoyed it and thought I would spend longer in South Carolina. Approaching Tony, I told him I thought I might be interested in staying on. He said, "No problem, stay as long as you want". Looking back this seems quite bizarre, with the fights for tenure going on here all the time – thank goodness for Sputnik – because of which physics was well funded for a while – the Russians will never know what they did for science in this country.

7.09 TEACHING IN SOUTH CAROLINA

University teaching in South Carolina is very different from England or Australia, and it has changed drastically over the years. I well remember my first class – the girls had the high, bouffant hairdos, and were exquisitely made up. I was very impressed. Of course, it turned out this was not on my behalf, but to attract the male students, because this was the prime environment for the mating game.

For me, the dramatic change was the miniskirt. It occurred while I was away for the summer, and arrived back the first day of class. The previous dress code was a skirt which came two inches below the knee, bobby socks and penny loafers. As I came into class, I found I could see the girls well above the knee. I found this most distracting, and had the greatest difficulty keeping my mind on the course. Of course I did not object to this however. The standard dress then became jeans and T shirt for both boys and girls. But there is still conformity. You have to have designer jeans, and Tommy Hilfiger T-shirts. To me, this seems to defeat the objective of such clothes, but then, I am an old curmudgeon. More recently however, it is a very short skirt or shorts for the girls.

The students came ill-prepared for the university courses. The training I got at school in England was very demanding. In America, it appeared there was little incentive to work hard. You would get into university anyway, if you wanted to. So, you could probably graduate from an American University immediately on leaving a British High School. On the other hand, after taking their many postgraduate courses for the PhD, American students might very well be ahead of the game at the graduate level compared with the British. In England, at least at Oxford or Cambridge, no courses were demanded after your bachelor's degree. All you required was a suitable research project, a thesis, and to pass an oral exam.

7.09.1 GRADUATE PROGRAM

When I arrived at USC they had no doctoral program in physics. This was instituted my first year. Tony French applied to the NSF and got funded to acquire two graduate students, Wieler Hurren, and Karl Fritz. We had to organize a graduate program – lectures on quantum mechanics and high energy physics for example. Hurren later became head of physics at the Citadel in Charleston. After graduating, Fritz later obtained a medical degree. The graduate program has grown immensely, so that there are now many students in theoretical physics, as well as experimental work in condensed matter, high energy and intermediate energy physics and sundry other fields.

7.10 STUDENTS' VAGARIES

I recall one unfortunate episode. I had given a lecture to a first year co-ed class, which had broken up into laughter at one point for no obvious reason. I continued with the lecture as if nothing had happened, but I was baffled. So I asked a friend of mine, Jap Memory. He said, "Well, go over what you said at the point where they broke up". I replied, "I said nothing. I had just made a mistake on the board, and remarked, "Well, blow me!" (a very mild expletive in England), "I should have put an x here." Jasper explained my boo-boo to me amid much hilarity, and I have never used that expression again in class.

We had some oddball students. One (was it Duke Torbert?) had a habit of falling asleep in class. One day he fell asleep in Ernst Breitenberger's class. Ernst did not awaken him at the end of his class, and Jasper Memory came in and started to lecture, so Duke woke up half way through Jasper's class, completely confused.

Determined, he later got a large cup of strong coffee. I think it was in my class. Anyway, half way though the class there was the most immense crash. Duke had fallen asleep and dropped the cup!

Then, my assistant told me that in a large (300) class I had, there was a student at the back who regularly had a can of beer each morning. I didn't know how to attack this problem, so did nothing. However, the last day of class I happened to be in the elevator with this student and his can of beer. "Why do you drink that?" I asked. "I don't have time for a regular breakfast," he said. Talk about a non sequitur! We have had some strange students over the years. There was an instructor of mine who was caught having sex with a student on top of the lab table. He went.

Father Reddy was an excellent student. In one experiment he was measuring the speed of light using a rotating mirror. Anyway, the image was exceptionally difficult to see, and without thinking, I said "Examine it carefully, you must look at it with the eye of faith," then it hit me – he was a priest! But he took it in good part.

7.11 SEGREGATION

When I first arrived in America, I found segregation not particularly noticeable unless you were observant – for example the white and colored drinking fountains – but I almost never use drinking fountains! My wife and child were at a downtown Woolworths for lunch when a sit-in started, which was very interesting. When they realized what was going on they left – in fact they were encouraged to leave I believe. When I arrived in the South I wasn't aware of bus segregation – whites at the front – and I am sure I sat at the back, but nobody said anything – probably my British accent protected me. I have a feeling that the bus drivers, for the most part, didn't want to make an issue of it unless a point was being made.

The university was completely segregated at that time – and, judging from the remarks of some of my colleagues, likely to stay that way. To my horror, there were people on the faculty who had gone to USC because it was segregated – just as there were people who would not be caught dead there for the same reason. To me, it was more a matter of interest to see how such an obviously unethical system could work. It did, with creaks and groans, although I was glad when that era was over, which occurred shortly after I had moved to Yale University.

I gave a talk about it on British BBC television in Manchester when USC integrated, because I happened to be in England at the time. The very same studio on Piccadilly in Manchester, where I had sung in a choir as a boy was now a TV studio. I recalled I had met Wilfred Pickles, the announcer with a ripe Northern accent, who was later bumped for someone with a "BBC" accent. One nice civilized thing about British television – before the live interview, they provided a glass of sherry. This both relaxed me and put me in the right frame of mind. They would never do that in America.

I had always wondered what it must feel like to be the first African Americans at USC. I realized when I went to meet with a friend who taught at the South Carolina State University which was then completely segregated – black-only! We had been discussing for an hour or so, when the bell went signifying lunch. We went outside talking, and I looked out on this vast sea of black faces going to and fro, all going to lunch too. It gave me the most peculiar feeling mostly of strangeness, not fear or anything. It suddenly struck me that this is how the African Americans who first came to USC must feel. Now we are fully integrated – perhaps!

7.12 CRAPPER

The American Physical Society held regional meetings. We belonged to the Southeastern Section and a regional meeting was held at Clemson University. I gave a paper on surface tension waves or ripples, an interesting topic requiring little experimental equipment, so very suitable for research at USC at that time. My first slide was of the motion or orbit of the water particles in the ripple, and it had exactly the same shape as a toilet seat. Elliptical, squashed a little on one side. Unfortunately, I had put the author of this orbit directly beneath. It was "Crapper"! As I showed this, there was a stunned silence – then hilarity. It took a long while to live this down. It is difficult to believe, but as a naive Englishman, I had no idea of the relation between "crapper" and "toilet". I did not make that mistake again.

7.13 NUT PAPERS – CYLINDRICAL WAVES

My interest in surface tension waves or ripples led to an investigation of solitary waves. The non-linearity of the equation describing water waves leads to the possibility of solitary waves instead of whole groups of waves. A solitary wave consists of one peak (or trough) which travels along on its own. The first investigation of these came from the work of Scott Russell, who followed the wave generated by a canal boat in Scotland for several miles in the late nineteenth century. I was particularly interested in negative solitary ripples. Absolutely no-one else was interested in solitary waves of any kind then, so my papers at the New York meetings were relegated to Saturday mornings, with the nut papers. These were papers from people who had devised antigravity machines, or believed that the fundamental unit of mass for all matter was that of the electron. Some of these were genuinely bonkers, but others, such as mine, were too far ahead of the game for anyone to realize they were important. I met some very interesting people this way. However, by the time solitary waves became important many years later, I had lost what interest I had, so I never did benefit.

The experimental work I did was with an undergraduate named George Walters – it was very simple, requiring primarily a large oblong fish tank, and a metal semi-cylinder.

7.14 PARIS

I took a trip to Europe, and had to visit Orsay to see their electron accelerator. Taking the usual overnight flight from New York, I arrived in Paris feeling lousy. I had this Arthur Frommer book called "Europe On Five Dollars a Day" and went to a recommended establishment on the Rive Gauch, the bank of the river Seine. It proved to be a very good choice, called the "Home Latin" if I recall rightly. The concierge was a great help and it was comfortable, and completely French, even to the extent of having a bidet (an item I had never seen before) in the middle of the room. I got completely soaked trying to use the damn thing. I then had to get out to Orsay, and discovered I must take a train from the Gare du Luxemburg. I spoke to the ticket lady, and she answered volubly and very rapidly in French, that I should take a certain train, and it would take me straight there. I boarded, the train took off, and I promptly fell asleep, to awaken at a station with the name "ROBINSON" in large letters. How had I managed to return suddenly to England? This was the terminus. I got off to find I had taken a train which branched off onto a sidetrack after a short distance, and the station was named after an English restaurant owner whose establishment had been at that spot many years ago. It seems that English cuisine is more appreciated in France than England!

I crossed the track, re-boarded, got off on the main line and took the next train out. This is where I made my big mistake – I went to sleep again. This time I awoke with the train stationary in the middle of nowhere, with not a soul on board. It was like one of those horror movies. I walked up and down the train – no-one. About ten minutes later, I saw an individual approaching along the track. It turned out it was the driver, and he had parked the train on a siding while he went and had lunch. He had not noticed me on the train, fast asleep. Back to the station, and re-board the correct train. This time I stayed awake.

I felt I must go up the Eiffel Tower, but it was closed above the lowest level which was a restaurant. So I had dinner there. Adjacent to me were three very attractive girls who were talking about Auburn University, which they were attending. I introduced myself as another Southerner, and we had quite a conversation. Then they said they had to get back to their hotel, but were a bit worried about doing so as they had no French. Since it was near my hotel, I volunteered to help them, and we took a taxi. There were two gendarmes chatting on leaving the taxi, so I asked them the way. They viewed me with these three pretty girls, and gave me directions. As they turned back I heard them say, "Ah, les Americains!" Unfortunately, that is the end of the story. I saw them to the hotel, and went to mine.

7.15 ACOUSTICS & OPTICS, ANECHOIC CHAMBER, COMPOSITE VIOLIN & GUITAR, CARBON FIBERS, SUN DOGS

Since I had agreed to teach a course on physics for artists and musicians – "Physics in the Arts" – basically composed of light and sound, I took a great deal of interest in acoustics, particularly reverberation time, and the structure of musical instruments. This led to constructing an anechoic chamber out of four by eight foot sheets of plywood, and highly absorbent acoustic tile, a topic on which I gave a paper before the acoustical society. An anechoic chamber is a room where any sound is immediately absorbed. I also constructed a "seventh power horn" on which an honors student wrote his thesis. He became a lecturer at Cambridge University in England. This horn was thought to be impossible to construct, but in fact it proved quite difficult to play. My acoustic interests got me in touch with Tom Rossing, who was into bells, gongs etc., particularly the then recently discovered ancient Chinese bells He has written much on bells, and my bell-ringing antics interested him.

A member of our engineering department had constructed a guitar top, and also a violin top out of a composite material. This is a substance composed of aligned carbon filaments, held together by epoxy resin. Regular violin tops, being made of wood, are each one different, and must be hand-crafted by tapping to get the right tone. The idea was that if you could easily make an ideal violin from the composite, you could make a hundred or more. Wood is elastically very anisotropic, but the composite could be made to have identical properties to any wood. Anyway, we took the violin and guitar to an acoustical society meeting, and they were well received. There was a famous guitarist who took the guitar away to practice. Some weeks later we heard from him. He had been practicing on the guitar, and said it had much improved. Now this was a problem, because once the top had cured, the plastic should not change at all (unlike wood). We finally worked out it was not the plastic but the performer that had changed. You get used to an instrument, and he felt, as he got used to the new guitar, that it had improved.

Anyway, the engineer got fed up and went off to India, and I never saw him again.

Atmospheric optical phenomena and optical illusions interested me too, leading to my "Physics in the Arts" course – so I wrote an article on sun dogs. These are two bright spots of light one can see a few degrees either side the sun. They are produced by ice crystals as they float horizontally.

Another interesting phenomenon was Haidinger's brushes. The polarization of light can be detected with the naked eye. The retina is "dichroic" – that is to say, the substance of which it is composed reacts differently to different directions of polarization. Very few people are aware of this, but if you look at a white sheet of paper through polaroid sunglasses to polarize the light, you will see a faint colored figure wherever you look on the paper. It appears to be an inch or less, and is yellowish horizontally, and bluish vertically. It is difficult to see, and I generally had only about 50% of my class actually see it. No doubt our ancient ancestors found their way about using this effect, because the light from the blue sky is polarized, so you can use polarization like a compass to find direction whenever the sun is hidden by cloud.

7.16 GATHERINGS FOR GARDNER – ETERNAL STAIRCASE AND HAIDINGER

Optical illusions are of importance to artists and hence formed part of my course. This led me to develop extensions of Penrose's "eternal staircase", immortalized by Escher. Penrose continued to develop the illusion so that it is not merely globally ambiguous, but also locally, so you can go round the staircase several times in different orientations before arriving back where you started. Unfortunately both Escher, (and Penrose's father who helped develop the original "impossible staircase") had died, so this particular illusion has never become famous. I extended it further into three dimensions, so it had no edges, and one version also took the shape of a cube. I discussed this a number of times with Penrose, and it led to my attending a famous "Gathering for Gardner" a collection of eccentric individuals who meet every two years in Atlanta. I went to one in 2002 and was astounded at the variety of projects. The people interested in many of these had no other outlet. I had lunch with the Englishman, Adrian Fisher, who spoke about mirror (and other) mazes. He headed a company which designed mazes, a fascinating topic, but there are no societies of maze designers – or at least, none of which I am aware. He showed everything from Hampton Court to the corn mazes in America, cut from corn stalks in the field to form a maze – different every year.

The Gathering for Martin Gardner is held every two years. I was invited and attended the fifth such gathering G5, but missed G6, attending G7

March 16–19 2006. I submitted a paper on Haidinger's brushes, using circularly polarized light. I have never been convinced about this, and feel it needs more work.

As usual, the program was eclectic. John Conway, the inventor of the game "Life", and Roger Penrose both spoke.

7.17 BILL SAVAGE AND TOM ROSSING.

My acoustical interests led me to become acquainted with Bill Savage and Tom Rossing. The former was a professor at the University of Ohio, where he organized an acoustics meeting. I arrived in early spring – warm in Columbia, freezing in Ohio – I had to borrow the taxi driver's coat to get to the hotel since I had not brought anything warm.

I first met Tom Rossing when he applied for a slot in the NSF workshop on teaching acoustics which I ran in the early '70s. It was when he had first become really interested, and he did very well – he calculated the effects of an inhomogeneous string for a violin string using math normally employed for transmission lines. He also almost won the state tennis championship. Our friendship continued through the years, fueled by an interest in bells. He became the guru of bells, gongs and drums. Prior to him, little had been done on the physics of these instruments, but he wrote numerous articles and books on the subject, with the result that when the Chinese dug up a set of antique bells, he went over to study their acoustical properties. These proved to be quite unusual. I gave a talk at his university the University of Northern Illinois, demonstrating the levitation magnet, held in place by gyroscopic forces.

7.18 COSMIC RAY NEUTRONS

When I arrived at the University of South Carolina, we had no research grants or equipment, and the question was, how should we start? Amongst other things, I had been working on cosmic ray neutrons in Australia, employing the miserable Szilard Chalmers reaction in permanganate solution. I swore never to do this again, but counters employing BF^3 were now easily available, which, though not as sensitive, were quite reliable.

7.18.1 WIS TV TOWER & SKIP HENDRICK'S RELIEF

The WIS TV broadcasting company erected a tall tower near Columbia – about 1500 ft high I believe – tallest east of the Mississippi then. Somehow I got permission to use this tower, and so we mounted our cosmic ray neutron detectors at the 1300 ft. level. Now climbing a 1,500 foot tower was not my idea of what a research physicist should be involved in. However, luckily the tower had an elevator. It was a very crummy elevator – but it worked – at least most of the time. However, I recall on one occasion, I went up with the engineer, a photographer, and a fourth person. When we were about two thirds the way up, the elevator stopped with a jerk. What to do? I personally was not going to climb down a ladder 1,000 feet high, although one was available. Neither were my mates. For a while we yelled, hoping someone would hear. However, it is a physical fact that the inverse square law reduces sound by a surprising amount with distance, so we went unheard. I was in the cab with the photographer, while the engineer and the fourth individual were on top. The photographer told me that the last time he was in the cab the rope broke, so the vehicle was in free fall. There was an emergency button, but he said he did not want to press it till they got near the bottom, because he would have to climb down the ladder. Anyway, he got out of it alive. To me, this sounded like one of the tallest tales I had ever heard, (even taller than the tower!) so after we got down, I asked the engineer in the control room whether he had heard of it. He said, oh yes it really happened.

After some considerable time, the engineer in the control cabin came out, saw what had happened, and reset the breaker. We got to the top in good shape. However, I am no great one for heights, so each time we went up, after reaching the small platform at the high level, I would get out and grip the railing for ten minutes. The floor of the platform was composed of expanded steel, so one could look at the ground a thousand feet below through one's feet. After recovering my composure, I continued with setting up the counter detectors. The counters were set running, and I descended the tower. At a later date I had a graduate student (Skip Hendrick) doing the experimental runs. He informed me much later about one problem. He would be up the tower for several hours during the runs, and of course there was no toilet. As he relieved himself over the edge of the tower, he wondered whether it would all evaporate before hitting the ground. I suppose it would depend on the relative humidity. The tower was an unusual asset at the time, and I remember they had a "Miss Tall Tower" beauty competition. The winner got to go up the tower, and I was there when she did so. Somewhat of an anticlimax I imagine, because there was nothing much except trees to be seen from the top of the tower.

We ran the equipment off batteries, but it would clearly be an advantage to use regular current. A power outlet was 50 ft. below us, but I was clearly not going to shin down the tower at that altitude. Luckily, a colleague of ours, Fred Giles, had helped pay his way through college by painting the gold balls on the top of flag poles. When a ball had lost its glitter, he would shin up the pole with a can of gold paint and a brush, and paint it. This was the ideal preparation for someone to climb down our tower and plug in the equipment, which he did.

7.18.2 BALLOONS

Sometime later, we were forbidden the tower for one reason or another. I believe it was because of problems with insurance, in case something happened to us. I tried three other detection techniques. The first was a balloon. We inflated a gigantic – (20 ft. or so) weather balloon, and attached the counters to it. Since we had little money, it was inflated with hydrogen rather than helium. I was somewhat scared of this balloon, in case it exploded, so we decided to apply a match to see what would happen. The balloon did not explode – it burned vertically upward with an immense "woomf" – but no explosion. Anyway, we sent up several balloons, tethered of course, at Fort Jackson. This military establishment was not under the jurisdiction of the FAA, so we did not need special permission for the ascension.

It always seemed windy whenever the balloon went up, and finally one broke away. We were accompanied by a Major in the army, so I said to him "shoot it down, shoot it down!". He replied, "Do you know how difficult it would be to hit that thing at a distance of a mile or two?" Anyway, he phoned the FAA, and that was the last we heard of it. They probably thought it was some scatterbrained idea of the army to bring down planes.

7.18.3 BIG KITES

Since we had such trouble with wind, it seemed natural to try a kite. We built a gigantic 30 foot kite of brilliant red material (Prof. Luther Barre, who bought it, said this was the cheapest fabric). We flew it down by the Congaree river, where there was a suitable open space, and in fact did get our equipment up using a big winch. Unfortunately, the undergraduate who was helping us, let go of the winch handle. It came around hard on his knuckles, and he could not stop it – he ended up in the hospital. People on the other side of the river, viewing this big red object flying in the sky, opined it must be a UFO, as they later informed the local paper. We did not disillusion them. I have always been suspicious of UFO reports ever since.

7.18.4 AIRCRAFT

The fourth method we tried was a conventional plane. The Air National Guard had in their possession a DC3, or C47 as it is sometimes called, (similar to the one I employed in Australia). It was later overturned in a hurricane. General MacIntyre gave us permission to use this, so we mounted our equipment and off we went. I was interested in low altitudes, so we flew from 100 to 1000 feet above lake Marion. Flying at low altitude in the summer is a miserable business because of the strong thermals. I rapidly got air sick, but luckily, the student with me (Tony deLoach) did not. I found the best place was the cockpit. If the plane drops a little, the tail must drop further, to bring the plane back up again, so the tail waggles up and down a lot – not so the cockpit. I let the student do the work, and this way survived an unpleasant experience.

7.18.5 EXPERIMENTS UNDERGROUND

One part of the experiment investigated the cosmic ray neutron flux in the earth as a function of depth. We cut a narrow slot about one meter deep, and long enough to take the neutron counter and lined it with a metal container Then we measured the flux descending this hole. A year or two later, the area behind the museum where we had dug this hole was landscaped into little hillocks, so the slot is probably still there somewhere underneath.

7.19 NEUTRON GENERATOR BUILDING

The cosmic ray work was interesting, but I felt we needed more experimental research. So I wrote a proposal to erect a building in which I could do nuclear physics. We obtained the accelerator with funds from the National Science Foundation. The machine was a 400 kev High Voltage Co. Van de Graaff accelerator, which was to be used to bombard deuterium and tritium targets with deuterons, thus giving a profuse yield of neutrons which could be used to investigate (n, 2n) and similar nuclear reactions. ^{3}He was also used to bombard, since few reactions had been done with it, and it had a low binding energy.

The building had very thick walls – more than four feet thick of poured concrete with rocks in it – which provided strong shielding for the experimenters from the neutrons, both low and high energy, but it had a thin roof, with the idea the neutrons would exit through the roof and vanish.

One interesting feature is a metal plaque on the outside of the building near the front door. It has on it the state legislators involved at the time of this project, one of whom is listed as Earnest F. Hollings. I wonder if Hollings has ever realized the "Importance of being Earnest"?

The building has always been referred to as the "neutron generator building", because of its original purpose. In fact, it has practically never been used for neutron production. This type of work was done at Duke University, and instead we did a lot of ^{3}He reactions, and later, solid state research. This is perhaps just as well, because shortly after the building was up and we started research, they decided to erect eight story structures around the NG building. Because of the thin roof of this building, anyone in the new buildings would have received a nice dose of neutrons.

It was the practice not to air condition university buildings at that time, which made working in the summer abominable. However, I had cleverly demanded my building should be fully air conditioned, ostensibly because otherwise the delicate equipment would be ruined. The result was, that first summer which was very hot, all the faculty, both theoretical and experimental, moved into the air conditioned NG building, an unanticipated event, but much appreciated by all.

Much of the work involved in getting the building up was done by Ed Lerner, because I was away for the summer at Oak Ridge Laboratory. I had requested a red light go on over each door in the building when the machine was running. Instead of which, the architects had arranged that, when you turned on a light in a room, a red light went on on the other side of the door. Whether they thought we were setting up a house of ill repute or not we shall never know. Over the years, this building has served a number of different purposes, but the thick walls will make it very difficult to demolish. One curious use was to execute rats for medical purposes. For some reason, there were edicts against executing rats, used for research, in the main physical science building, but no such objections were raised to doing so in our building, so that is what happened. The bloodstains from the guillotine are still visible on the walls.

I had a concrete pier inserted to some depth in the floor when the building was erected, isolating it from vibration for seismic measurements – but it has never been used for this purpose. One cannot predict the future.

7.20   HYDROMAGNETIC WAVES IN THE IONOSPHERE.

One of the strangest experiments I became involved in was the investigation of hydromagnetic waves in the ionosphere. The ionosphere is the conducting layer above the lower atmosphere. Bursts of solar radiation falling on this trigger ionic waves having a certain mathematical similarity to ocean waves, but high up in the ionosphere. They travel quite slowly, relative to the speed of light, and may be detected from the ground.

We got into this because the army Signal Corps was looking for a suitable site on the east coast – they had one way up north in Vermont or New Hampshire, one in New Jersey, and they needed a southern site. We set up our detectors at Fort Jackson, which was an ideal place, because we were bothered only by bombs and artillery – but not by electronic noise. The reason behind all this was that an atom bomb was to be exploded in the ionophere over Johnston Island in the Pacific, and it would trigger some large hydromagnetic waves. Since this was likely to be a one-off thing, because of environmental objections, we wanted to make the best of these observations.

We started off experimentally with some coils six feet in diameter, lent us by Cal Tech, who had developed equipment for seismic detection. These were not all that successful, so finally we constructed a single coil of eight turns or so, which completely surrounded Fort Jackson. Since there was no electrical gear on the firing range this worked fine. Except that the coil kept getting struck by lightning in summer, as it was hung from the trees. As a result, a narrow trench was dug right round the fort, many miles long, and the coil laid in it. It still got struck now and then, but rarely. The ground around the struck section melted to give tubes of fused sand. Our results were quite interesting, since we let the equipment run continuously. We detected "pearl oscillations" for which the print out resembles a string of pearls, but finally the day approached when the bomb would be exploded. This presented a problem, because we were not cleared to be told when the bomb would be let off. Nevertheless, we needed to know, in order that our equipment should be in good working condition. As a result, a system was developed. The people at the hexagon (Signal Corps headquarters) would phone us and say "something is going to happen tonight" but they would not say what. Next day, when the bomb was not exploded they would say "we couldn't let the bomb off last night – bad weather". This would go on for a week or two, and then the bomb finally went off – and we got a good record.

We had to wander over the firing range, in order to fix things, and we checked ahead of time that nobody was firing. However, one time Rudy Gaedke, the graduate student assigned to the experiment, had to go out on the range. He had found his Texas outfit – cowboy boots ten gallon hat, etc., was ideal for the hot, sandy countryside. The recruits on the range, getting all ready to shoot, looked up to see this Texas cowboy suddenly appearing over the hill in full gear. Gave them quite a surprise.

7.21 OAK RIDGE NATIONAL LAB

Not having any research funds the first year, I applied for and got a summer appointment at Oak Ridge National Laboratory. This was a big advantage – I got paid for the summer, and also did work at one of the best labs in the country. The trip to Oak Ridge proved most interesting. There were then no interstates over the mountains, so we took the two lane road over Caesar's Head in South Carolina, and spent the night in Cherokee. I was being paid the princely salary of $6300 a year, so we had little money to play with. My wife talked the motel owner to let us (wife, child and myself) say in a garret (small, but clean) for $5 a night!

7.21.1 THE CAS WALKER SHOW & THE BODY IN THE ROAD

Anybody who lived in eastern Tennessee in the late 50s will remember the Cas Walker show. Cas owned a series of supermarkets in this area, and ran an early morning TV show. Now, you have to remember that TV today is not the same as it was then. When I say he ran the TV show, I mean he really *ran* it – he was always there himself live at six or seven in the morning, together with all the best Country and Western stars, in person, and chatting with Cas in a friendly fashion, together with vocalizing their latest releases. It was an eye opener to somebody bred on the BBC. Then, there was the occasion I came in to work to find our assistant quite distraught. Driving in, he had just discovered a body in the road, shot to death. Of course he called the sheriff, who, on observing the body, said "Oh, that's old so and so. Must be such and such shot him", and went off to find the culprit. Feuds were still the vogue in Tennessee at that time.

I used to travel to work with a guy who had just bought a new Ford Thunderbird. The old road (it has since been replaced) did not follow the contour of the land, and had some tremendous bumps. My friend found, if he accelerated enough before one such hillock, the car would take off, and have all four wheels in the air, feeling like a roller coaster.

7.21.2 MT. LECONTE AND THE BEARS

In getting to and fro to ORNL, one had to traverse the Blue Ridge Mountains – before the interstate was complete. Many interesting things occurred on these trips through the Smokies. I had always wanted to climb Mount LeConte from Alum Caves Bluff. The mountain name itself is interesting. There were two LeConte brothers who were professors at the University of South Carolina before and shortly after the Civil War. One was a geologist, and Mount LeConte is named after him. The interesting part is that after the Civil War the brothers went out to California, and one was the first president of the University of California at Berkeley. So there are LeConte buildings at both the University of South Carolina, and Berkeley. Not only that, there is another mountain in California named Mount LeConte after one of the brothers.

After a conference at Oak Ridge I phoned the lodge at the top of the mountain on the off chance they had room – mostly it is booked up – but they actually had space, so I reserved it. Unfortunately, all I had to wear was my suite, for the conference. There is no transportation to the lodge – you have to climb up there along the path. I parked and set out from the bluffs wearing the suite, and with my pajamas in a bag I had in the car. Since it was showery, I took my umbrella. I had gone but a short distance along the trail when I met a black bear coming towards me. The path had bushes either side, making it impossible to avoid the bear. Not wishing to confront it (it was a big bear) I turned and reversed my path. The bear followed. I crossed a log bridge, but so did the bear. Finally, I got somewhat in front of the bear and reached a fork in the path. A while later I looked back – no bear. I started up the mountain again – thank goodness the bear must have taken the other path. Continuing up past the caves, I heard a noise in the bushes – it was another bear grunting and making a fuss! Luckily it was not interested in me

— so I crept past and went on my way. Then, to my surprise I ran into yet a third bear. It had found something nice in a trash can of some kind, and was fully occupied.

It started to rain, but was very hot and humid, so I arrived at the lodge wearing my suite pants, naked from the waist up, and with my umbrella up. As I walked in the door of the lodge, the man at the desk looked up and said, "You must be the Englishman".

I slept in a tiny log cabin — just large enough for one bed, and with a wood stove in the corner. I saw the most magnificent sunset before retiring.

The evening had been spent with a very congenial bunch of people—there is something which brings out the best if you have just climbed a mountain, so there was much singing and telling of tales.

I used to go skiing (very badly) at Cataloochee and Gatlinburg on the way over. And skating too, at which I was even worse, having learned when I was well on into my thirties, at a small ice rink in Columbia.

7.21.3 RESEARCH AND LIFE AT OAK RIDGE

Since I had no clearance, I was not allowed in the main body of the lab, but the Van de Graaff accelerators were outside the fence, so I had no problem doing my research, albeit, the release of radioactive tritium in the accelerator building had complicated the situation. An accidental escape of this gas from the equipment had occurred, and the air conditioning system helped distribute it.

I had to bring my lunch, the restaurant being in the main building. My clearance demanded I take a medical test, where they found out I had had tuberculosis, a fact of which I was totally unaware. I suppose I got it as a child in England, since our local farm cows were not tuberculin-tested.

7.21.4 THE REVIEW PANEL

As an outsider I was invited to attend the review panel meetings each section of the lab had once a year. I found these most illuminating. We always had a big evening party, often at the local country club. At that time, Tennessee was still completely dry, and a dry party is no fun. The country club would, however, lay on liquor, and all went well — except it was quite illegal, and to see bigwigs, Eugene Wigner, etc., with glass in hand came as quite a surprise. Anyway, the lab liked to have pictures of all these people busily discussing physics (which in fact they were) but they could not be seen drinking. So someone was assigned to remove everyone's glass when a picture was taken, and they would be restored to the correct person immediately afterwards. It must have been a problem remembering whose glass was whose.

7.22 THE CENTRAL CORRECTIONAL INSTITUTION

I was once asked to give a talk to the inmates of the jail here in Columbia, the Central Correctional Institution as it is called. I had said I would talk about light and waves, and I think the organizers and I were to some extent at cross purposes. Anyway, I prepared a talk about waves and how light works in simple terms, with lots of demonstrations. It was very interesting entering the building – the only time I was there I am glad to say. There was lots of searching and opening and closing of iron gates, but the staff were very pleasant. On the whole, the talk went well. There is something about having a captive audience. However, the experiment which made the day was a large hologram I had brought. This was in the days when holograms were a new item, and this one was a large transparent picture of a pretty girl, in three dimensions of course. As you walked past, she blew a kiss at you. I had forgotten the imprisoned inmates did not see too many women, and those they did see were generally not that attractive. As a result, after I had shown off this hologram, they lined up and proceeded to parade in front of the hologram several times. There was no doubt that the hologram was the hit of the talk!

My dad and me at age 10

Me and my PhD outfit from Cambridge!

1 MeV Philips stack for accelerator. This shows the stack of Greinecker doubler devices which provided the 1 MeV to the accelerator. It had a tendency to misbehave, especially if one of the rectifier tubes went wrong!

The oxygen mask we had to use at high altitudes in the DC3 - no pressure or heat. This experiment was performed in a DC3 (C45, Dakota) flying at 20,000 feet over Australia. We had a box containing a beer barrel of sodium permanganate solution neutron detector. A drop spilling on the wooden floor would catch fire, which we had to extinguish before the pilot saw it.

Christopher, me, Margaret (my wife) and Christian (her brother). Margaret's family escaped from Czechoslovakia the first year we were in the USA. Margaret had never met her brother, who was now 21 years old. He visited us in about 1968 - and this was taken as he got off the train from New York.

Our group ready to load the equipment onto Coke's 1936 Lockheed electra. This plane proved invaluable in transporting our technical equipment quickly to SREL, so we could occupy running time which would otherwise be lost.

Learning to fly in Cessna 150 (or172)

Ferry to Aarhus

Eating Ginger Bread Man in Aarhus

Me Late 50's

I have always wanted to be an actor, but at school the best I could do was one of the people who carried off the bodies in Macbeth. My profession as a teacher has led me to have a powerful voice, essential for a good actor, but I have a terrible memory, which foreclosed on an acting career. Imagine my surprise then when I received a phone call at eight in the evening some years ago inviting me to take part in a production of "Appointment with Death ". Some years later, I got another call to be the burglar (again, British) in Michael Frayn's "Noises off". This is the one play I wanted to be in. This photo is a cast picture of the production.

Me and my accordion. I was about sixty-five when I saw this accordion in a second hand store in Hendersonville NC. It attracted me, so I bought it - for $60. I took up playing it - never had a lesson, so I (as do most accordionists) play by ear. However, the instrument is designed for people like me. The accordion is a happy instrument. It cheers people up when they are down - and you can play it on your own - no backup is needed, though it helps. A colleague sold me a 120 bass instrument (the left hand keyboard has 120 chords - a standard accordion) that his kids had accidentally acquired - but it was heavy. I finally settled on a musette accordion, such as is played outside the Paris Bistros. This is not too heavy, and has all the reeds I need. I have had three such accordions. The first was stolen, the second melted. The reeds are attached by beeswax, and in the hot South Carolina weather, the wax melted, and the reeds fell off. I also have a huge accordion built like an organ I acquired at a sale. I have to get someone else to put it on me because it is so heavy. I like to busk. This entails sitting on the accordion box and playing at fairs and markets. Originally I did not collect money, but so many people said, "Where's your hat?" that now I put out a hat, and give the money to charity.

8 CALIFORNIA

8.01 SUMMER & FALL IN CALIFORNIA. CAL TECH PARTIES

As I mentioned, when I first went to USC, I had intended to leave after a year. At the end of the second year I decided to see California, and wrote to three universities on the west coast for a temporary position, Stanford, Cal Tech and Berkeley. I thought, if I was very lucky, I might get an offer from one of them. To my utter astonishment, I got offers from all three. Now, this might have a tendency to boost one's ego until you examined the situation. This was not long after the Sputnik business. Universities write proposals for funds, and generally there is a 50% or less chance of them being accepted. Immediately after sputnik, the government decided to spend more on research, with the result that research groups had more money than they knew what to do with – I may say, it has never been the same since. Anyway, I was a very viable and legitimate way to spend this money, hence the offers. I accepted at Stanford, and decided to spend the fall at Cal Tech. Berkeley I had to turn down.

8.02 CROSS COUNTRY LOCKJAW

I drove over to Stanford in our old Plymouth. The day I started, the radiator gave out, and I had to have it fixed. After this inauspicious beginning, I was very tight – very tight mentally, and at the end of a long day's driving, I found I could not open my mouth. It was like a mild form of lockjaw – but I could not open it. I had some difficulty reserving a motel room and eating. If you ever try reserving a motel room with your teeth stuck together – don't. I think the motel owner believed I was some sort of gangster – you know, they always talk with their mouths closed in the movies. Eating, too, with your mouth shut is a hazard. I managed a little soup. Anyway, the following day, after a good night's rest I was fine – but it got worse as the day wore on. At about six in the evening it was quite bad, and I stopped for a beer at a small bar somewhere. Much to my surprise, I was fine afterward – the lockjaw was due to tension, which the beer had relaxed. After that I always had a beer in the evening.

8.03 HELD BY THE POLICE

I drove across the most southern part of Illinois, somewhere near Cairo, I believe. No interstates then – all two lane roads. I visited a state park, and while I was waiting at a T junction, a car travelling along the main road suddenly applied the brakes, and skidded off the road into a plowed field. It must have been doing 70. I got out to see if all was OK, and appearing so, drove on. I had gone a good 80 miles, and was on the border with the next state when I was flagged by a patrolman. He had recognized me because I had a child's tricycle attached to the top of the car – as well as everything but the kitchen sink. He said the car owners wished to bring charges against me, and we had to go back. He was very pleasant, and we chatted on the way back. It turned out that the people in the car found they had no case to stand on so the patrol officer ran me back to my car. He pointed out that the reason for nabbing me was that, had I passed into the next state, they would have had great difficulty bringing charges against me.

I remember running out of gas in the middle of Kansas. Again, it was all two lane roads – no interstate. I was close to a farm, and walked back to the farm house, some distance from the road. The farmer was friendly, and let me have enough gas to get to the next filling station. He had his own supply. I wonder, would this still be true? I had to get a new set of shocks in Topeka Kansas, which took all my cash. What I had not realized, as a naive newcomer, was that in those days (no credit cards) banks were local, and you could not cash a local check (in my case, from South Carolina) anywhere but your home state. I was used to England, where you could cash a check anywhere. Credit cards have changed all that.

8.04 OUT OF CASH IN RENO

By the time I reached Reno, I was completely out of money. If you have ever tried cashing a check in Reno, believe me, drawing blood from a stone is easier. It is one of the few times where my British heritage proved useful. One motel owner had married an English girl he met when in England during World War II. Luckily, their marriage had been successful, and he let me pay for the motel room with a check. I passed through Virginia City on the way, and stopped at the building Mark Twain had occupied as a journalist. Not wishing to buy anything, but to give the impression I might, I put a quarter in a slot machine. Oddly enough I won, and the machine delivered a whole bunch of quarters with much noise and ringing of bells – very welcome. To promote silver, all change at gas stations was given in silver dollars – great cumbersome heavy things – but – I wish I had kept them.

I was astounded at the border with California, on Lake Tahoe. They searched my car – for contraband fruit. The insect problem, I believe. Luckily I had none.

8.05 STANFORD

Arriving in California, I had no money at all. I went to a bank in Stanford with what struck me as the most peculiar name – the Anglo Crocker National Bank. The manager opened an account, and loaned me enough to get by for a week, while my funds were being transferred. Much slower in those days. I shall always look upon them with gratitude – I don't know what I would have done otherwise. Stanford was a good introduction to California. Almost no-one I met had been born there. It was a lovely place, but very expensive. I rapidly put away the idea of staying permanently, when I realized with a wife and child, I could not survive and enjoy life on an assistant professor's salary.

8.05.1 CARL BARBER, GERRY PETERSON, AND BARNEY GITTLEMAN

I ended up working for Carl Barber – a delightful man who had, if I recall, five daughters. He said with a wife and five daughters (no sons), if they all ganged up on him, it was impossible to refuse them anything. I worked with Gerry Peterson on the 180° scattering and electro-disintegration of a number of nuclei. We observed the giant magnetic dipole resonance. A guy from Argonne National Laboratory, near Chicago, (Dieter Kurath) did the theory.

In Stanford I befriended Barney Gittleman, who was working on the design of the first electron storage ring, associated with the big accelerator. I remember seeing the synchrotron radiation coming from this ring which was partly made of glass – I believe it was about the first time anyone had paid any attention to it.

Barney was the student of Gerry O'Neill, a Princeton faculty member who got interested in space travel, and developed the concept of a satellite composed of an inflated transparent plastic sphere, which could contain a complete ecosystem, so you could live there indefinitely. This became a popular item, causing Gerry to take to wearing a toupee, and becoming an incipient "hippy" in order to popularize it, and look attractive on television.

8.05.2 THE TWO MILE ACCELERATOR.

While at Stanford, "Pief" Panofsky headed the big accelerator project and I would visit his house in the evenings when he had a seminar. It was a beautiful place in the hills above Stanford, with a great view. One time Pief was in Washington, and his wife was looking after things. Just at the end of the seminar, Pief burst in, saying "We've got it" – congress had decided to appropriate the money for the two mile accelerator, which has been very successful. Strangely enough, I have never returned to Stanford to see this thing.

8.06 SANFRANCISCO WEEKENDS AND THE ANTS

San Francisco would be the ideal place to live, were it not so expensive (as my son Michael found recently). We would go down to the zoo if the weather was hot – the coast was pleasantly cool – or, were the weather cool, we went fisherman's wharf, or across the bay to Berkeley. From Stanford there was a lovely road over the mountains to Half Moon Bay (where a type of fish called grunion run at full moon, and people would take them from the surf and put them in a bucket) and Santa Cruz (a tiny village then with a carousel on the beach for the kids). On the way one time I stopped to take a picture looking back at Palo Alto, and a bunch of ants climbed up my leg. Have you ever tried to remove ants when you are in long pants? I took down my pants as quickly as possible, and beat off the ants. My son considered this hilarious. I did not.

8.07 BERKELEY

Whilst at Stanford I befriended another summer visitor – a man working for the Naval Research Lab. He administered grants, but had taken the summer off to do High Energy Physics research at Stanford. The Navy allowed this, and it was a good idea. Anyway, he had to visit the University of California at Berkeley, which had a grant for cosmic ray research from the navy, to observe how they were doing, and he invited me to come along too, which I did since I had never previously been there. The Berkeley people assumed I worked with him in the navy. I did not disillusion them. It is the only time I have ever been on the other side of the granting system. Otherwise I have always been trying to get grants, not distribute the funds. We were treated like royalty. Of course, observing their work was interesting, but even more so was the way they treated us. We went out to lunch at the best fish restaurant in Berkeley, down by the bay. It was delightful – but I am afraid, this proved a unique experience.

8.08 FROM SAN FRANCISCO TO LOS ANGELES

We drove from Stanford (Mountain View cherries and all) to Pasadena in our old hard top Plymouth, along the coast road – a very delightful drive, with beautiful views. However, driving along the cliff, we met a considerable belt of fog – so I could scarcely see the road, but heard the waves breaking hundreds of feet below. Luckily, my rather nervous wife was asleep during this period. We stopped at Monterey, and I hunted around to find the fish packing plant, and the Bear Flag restaurant mentioned by Steinbeck in his books. All had been torn down or done away with. It is interesting that so many people asked to see them that they had to rebuild them at a later time for the tourists. We looked at Pebble Beach golf course, but did not drive around. We spent the night at a a place called "Morro Bay". Then on past Big Sur and the mythical Hearst castle, which we saw from the road, to Los Angeles, getting lost in the complex overpass in the middle of town. We reached Pasadena, and stopped at an apartment for rent. As we stopped, the engine fell out of the car, so we decided to take the apartment – it proved a good idea.

8.09 WILLIE FOWLER

I went to Cal Tech at the invitation of Willie Fowler, though I worked mostly with Tommy Lauritsen's group. Perhaps invitation is too generous a word, since I had written to Cal Tech during my first year in South Carolina, and he had replied with an offer. At Cal Tech I saw little of Willie, but he was a remarkable personality. The last time I saw him he was telling a tale that when he was on his way to accept the Nobel Prize in Sweden, he ran into Barbara McLintock also from Cal Tech who had received the Nobel Prize (in biology) as well. He said, "You know, I have always wanted to kiss a Nobel Prize Winner". She said, "So have I". So they kissed!

Willie loved a party. We would meet at Los Alamos (many years later) in a collaboration at the pion factory there, and have a wild party, on arrival. Only problem was, Willie and the Cal Tech group would have come from the west coast, and we came from the east. So to him midnight at Los Alamos was only eleven o'clock and to us it was two in the morning. So, when we got tired and zombie-like, he and the west coast contingent were still raring to go.

8.10 TIJUANA

I took a trip to Tijuana whilst I was at Cal Tech, going by bus to San Diego, a beautiful city. I remember watching the navy flying boats doing touch and go landings in the harbor, while I was sitting on a lawn near the light house. I took a bus as far as the Mexican border, where it stopped. Everyone ran across the border and jumped into taxis, so I did the same, under the assumption they were going to the city. After I got in, the taxi took off like a shot from a gun, everyone talking in Spanish, of which I understood not one word. The driver went at great speed along the two lane highway, overtaking other vehicles on the grass verge, which I did not much enjoy. I later discovered one should beware of taxi drivers who have a statue of the virgin Mary on the dashboard. They believe they are protected. Anyway, I finally worked out we were not going to Tijuana at all, but a bull ring where the fight was imminent.

We arrived at the bullring by the sea, a new ring in fields on a cliff – and nothing else at all – no shops, no nothing. That is how I saw a bull fight by accident. I could not return to the city, because everyone was going out to the bull fight. I got a seat way in the back – sol, not sombra – so I saw little of the blood – but I enjoyed the ceremony, the mariachi type band, the toreadors, the colors – then I took the bus back to the town of Tijuana, where I bought a playable miniature violin for $2 – I still use it in lectures. I suppose a lot of Tijuana at that time would be described as "sleazy" – but it was still fascinating – the bright Mexican colors. The bars were definitely "interesting". There was also a very old bull ring in the town center. I walked across the border, and took the bus back to Los Angeles.

8.11 PHYSICS AT CAL TECH

The Kellogg radiation lab at Cal Tech had been built in the mid-thirties. At that time, Charlie Lauritsen had then emigrated from Denmark with his son Tommy. In those days, Los Angeles, and Pasadena were beautiful small cities. There was no smog, and life was peaceful.

Arriving at the lab, I had noted a large T embossed on the hillside. Tommie told me this stood for Throop – originally Cal Tech was Throop Institute, a Unitarian institution. Under Charlie Lauritsen's leadership, they had received a large grant from the Kellogg company to study insulators for high voltage transmission lines. Somehow, this got converted to high energy accelerators which were an MeV or two in those days. Anyway this was the beginning of the tradition in nuclear physics over there. There was no air conditioning then, so the lab had the most bizarre and inefficient system. It consisted of fans blowing air over wet blanket material pulled by rollers. The water evaporating cooled the air – but the smog was unaffected and appalling.

Nevertheless, they had the brightest students, and greatest faculty imaginable.

Some graduate students at Kellogg had been delayed in their thesis work by a problem with the accelerator they were using. They had set up a sort of club in their quarters at the lab, to pass time until the accelerator worked again. I also had a big problem. I was running on the tandem Van de Graaff accelerator one night when the vacuum in the accelerating tube failed. Bells rang, whistles blew, and the liquid nitrogen traps started to look like steam engines, blowing vast quantities of condensed moisture into the air. The previous users of the machine had not removed the roof shaped slits from the beam line, and the heat deposited had melted the support, allowing air into the accelerating tube. It took two weeks to fix, during which time my name was mud. However, I was not the only one with problems. Ward Whaling had gone into the vertical accelerator tank before it was fully emptied of the insulating gas, and passed out as a result of inhaling the gas. It was nonpoisonous, but asphyxiating. He rapidly recovered.

I was invited to give a talk at Cal Tech on my work at Stanford. This was one week after Rudolph Møssbauer had spoken on "How I won the Nobel Prize", which he had just been awarded. I felt rather like the manure cart which follows the Lord Mayor's coach.

The thing that most haunts me from my days at Cal Tech is my memory of the parties. I have never been to such enjoyable parties either before or since. One has a tendency to think of "wild" as an adjective – I would prefer the word "uninhibited". By later standards they were probably quite mild – after all, all we did was drink, dance and talk. Yet, under such circumstances, physicists are notoriously inhibited and we certainly were not. All who went to such parties recall them with a great deal of pleasure. All the best people attended – from Feynman on down – and we talked about everything. Ah, those were the days. It did not last. Very many years later I got invited to a meeting at Cal Tech. On the invitation it said, "We will have a party just like in the old days".

8.12 RICHARD FEYNMAN

We all had lunch together at the Athenaeum, a curious place possibly imitating the club of the same name in London. The Athenaeum was designed employing the "Southern California" style Spanish architecture, but paneled like an English country house inside. Feynman often lunched here. He had a most stimulating personality but was exceedingly conceited. He believed he was the best physicist in the world – the only problem being – he probably was! Nevertheless he was a fantastic conversationalist, everyone got involved, and it was always exciting, whatever the topic. There are a few people in this world who can make whatever you are doing – and whatever they are doing – sound interesting. Feynman was one such. Yakir Aharanov is another. However, sometimes it got too much – Feynman was notorious for asking seminar speakers simple questions which they nevertheless were unable to answer.

Lunching with Feynman is something I will never forget. After discussing and arguing with him while eating, I went away on a cloud. As I recall, he had been working on C^{14} and N^{14}, and I had just completed the electrodisintegration of N^{14} at Stanford which greatly interested him. I was at a talk one time where Tommy Lauritsen had to stand in for Charlie Barnes (a contemporary of mine at Cambridge) at a seminar, with Feynman sitting on the front row. Tommy wrote on the board, in large letters "DSMF", then walked up and down in front of it thoughtfully. Feynman, who could never keep quiet, said, "What does that mean Tommy?." "It means, 'don't shaft me Feynman' ," said Tommy, looking him in the eye. Tommy did not want to be distracted since he was unfamiliar with the material, and, in fact, Feynman did keep quiet for at least ten minutes.

After Feynman died, I received a short biographical sketch from Ted Welton, who had been Feynman's contemporary at Princeton, I believe. Having Feynman, who even then was incredibly brilliant, as a fellow student practically made Welton give up physics. Feynman was John Archibald Wheeler's graduate student, and probably Wheeler was the only thesis advisor who could have put up with him. Wheeler (who worked with Bohr on the possibility of the atom bomb) was a true Southern gentleman, very polite, but with a deep sense of humor. I recall discussing with him the three kinds of fundamental particles called quarks. Suddenly he stopped, thought for a minute and said, "It's like gasoline isn't it – comes in three flavors!"

I invited Welton down to speak at the University of South Carolina once or twice when he was at Oak Ridge. He suffered from narcolepsy – suddenly falling asleep. Strangely, this occurred when he got excited, not when he was bored. So, in the middle of a lecture he would put his head against the blackboard, take a nap, and then wake up suddenly and go on as if nothing had happened. One time I had to leave at dinner, and left friends in charge of him. Luckily I had warned them, because he dropped off in the middle of dinner. Another time, he came in to my office, and asked to take a nap fifteen minutes before his talk. He sat in a chair and immediately went to sleep. Fifteen minutes later he woke up, and gave an excellent lecture, without napping.

8.12a QUARTETS AT THE ATHENAEUM.

One of the interesting experiences when in residence at the Athenaeum was the string quartets and such which gave performances every now and then. These were quite marvelous, and I was puzzled how they managed to recruit such accomplished performers. Talking to one of them, it turned out that they were all employed by the movie studios to play for films being shot. This demanded fantastic accomplishment. The music had to fit perfectly with the action on the screen. The scene was shot first, then the performers played whilst watching the action. To help them, a line was drawn on the film itself, starting top left, finishing bottom right. As the film moved through the gate, one would see the line move from left to right across the screen. Anyway, developing such skills was a very profitable business – they were well paid, but doing this work on a permanent basis could be very dull and boring, and with no audience to applaud, not very rewarding as performers. So it was that someone at Cal Tech had invited them to perform for us, and we certainly gave them rousing applause – which they well deserved.

8.12b RELIGION IN LOS ANGELES

My stay in Los Angeles occurred long before the riots which occurred subsequently. Nevertheless, the incipient unrest was quite obviously there. This was brought out on a trip I took with other young Cal Tech visitors to a nearby Baptist church, which was primarily (in fact to me it seemed completely) black. This proved an unforgettable experience. The congregation was very hospitable, and made us feel at home, in spite of the varied and curious nature of their visitors, who came from all over the world. What struck me was the nature of the sermon. The preacher, a well-educated man, started off in low key, talking about Christ's life, but slowly he got his audience worked up, until they were saying "amen" and jumping up and down, promising how they were going to live better lives, and not get drunk and beat their wives, etc. Now this may seem to be something to be poo-pooed – but much to my surprise, I got worked up too, and although I didn't jump up and down so much, nevertheless I left with a very positive feeling which I had not anticipated. I felt cleansed, as they say. Talking with the minister after the sermon, he was saying how the members of his congregation were really segregated in a ghetto, and limited in the jobs for which their employers would accept them. I felt that this was like a powder keg, and of course, later it went off, with disastrous results.

8.13 THE ULCEROUS RETURN TO COLUMBIA

I found I was getting more and more sick as the semester wore on at Cal Tech – so eventually I consulted the quack (doctor) for the students, He enquired my symptoms, and said immediately, "Oh you have ulcers" – it turned out, that as doctor to the Cal Tech students, although a GP, he had become an expert at diagnosing ulcers, since this was such a common problem of the students there. If I were going to be sick, I concluded I would be better sick in Columbia, so I took a plane home. On the plane, I had a bottle of Maalox, and a carton of cream. On the hour I took a swig of Maalox, and on the half hour, of cream. This was the treatment for ulcers at that time. The guy in the next seat was very interested. My doctor in Columbia, Dr. Gause, packed me into hospital. X-rays showed I had two ulcers, one in the stomach, the other duodenal. The duodenal is rarely cancerous, but the stomach is. Anyway, he decided on the Maalox and cream treatment, and after a while I got better.

8.14 SUMMER AT CAL TECH

I spent the following summer at Cal Tech. Sleeping on the balcony of the Athenaeum was interesting. There were many cots put out, so I could say I slept with several Nobel Prize winners.

I returned via bus, stopping off at the Grand Canyon. The bus broke down in the wilds of Texas, in the middle of the night. As the cheapest mode of transportation, it had a large number of service men's wives, and infants. They cried more or less continuously through the night – not my most pleasant experience.

8.15 THE NEW YORK MEETING

Physicists had an annual meeting in New York at that time. Every January, I regularly traveled up to New York by rail. I got on the train at about seven or eight at night, and met all the physicists from Florida and Georgia who were going to the same meeting. We got together in the bar or club car. As we passed into North Carolina, the attendant would say "last call for drinks". NC had a dry regulation, so you had to wait till Virginia for a drink. We picked up the University of North Carolina, NC State and Duke contingents in Raleigh, and had a nice party. Finally I would go to bed in the "Roomette". This was a baffling device to an Englishman. You could sit in an armchair, which had a toilet under it if I remember right, or you could pull down the bed, which occupied the whole space, so you could not have the bed down and go to the toilet at the same time. Anyway, it was nice to wake up in the morning travelling at high speed through the snow covered countryside, whilst lying in bed and observing through the large window. To me, this has always been true luxury, and it is a great pity that over time, such meetings ceased.

I recall having to walk several blocks from my hotel (probably the 34th St. YMCA in fact) to the site of the annual dinner. I arrived late, because I passed a large fire several stories up in a high rise building, and was interested to see how the ladder trucks and other fire-fighting equipment worked. It turned out the only space available at the dinner was at the press table, which was run by Sam Goudsmit at that time. After I had sat down, someone asked me what had detained me. "Oh," I said "There's a big fire about a couple of blocks from here," and went on to describe it. As one man, the press corps rose and departed. This was obviously much more important than the physics talk about to be given. Goudsmit looked at me sadly.

It was always said that New York prostitutes took the week of the New York Meeting for their vacation, there being no profit to be had while the physicists were in town, as opposed, for example, to the medical practitioners, etc.!

8.16 FISH HOOKS

I was with Ed Lerner and Tony French one time, and we had hailed a cab to take us somewhere. Tony and I were still relatively new in New York, and we had agreed to let Ed pay as a native New Yorker, and reimburse him later. That way, he could deal with the cabby and the tip. So, at the end of the journey, we got out and Ed paid. At which point the cabby said, "They're all the same these English. Fish hooks in their pockets that's what they've got – fish hooks in their pockets."

8.17 THE NEW YORK COP AND BILLY GRAHAM

Another interesting experience occurred when several of us had arranged to share a room to save expenses. Joe Johnson was one of these, and he had invited a friend of his in, who was a New York policeman. We had ordered a fold away bed, and the servant arrived with the bed. Normally they wait for a tip, but just before this individual arrived, the cop had taken off his jacket, displaying a gun in a shoulder holster, which the police carry when off duty. The guy took one look at this and vanished! We went out to a movie in the African American district of 42 St. Whenever a black guy shot a white guy in the movie, the audience cheered. This display didn't faze the cop at all – he was used to it, but we cringed .

Joe had an interesting tale about a party when he was an undergraduate at Stonybrook. A friend had acquired one of those gigantic inflatable replicas of a liquor bottle. As they were about to leave for the party, a knock came at the door, and the friend answering it flung open the door saying, "I'm supersot". It turned out it was Billy Graham, there to pick up his child for whom one of the party members was baby-sitting. Graham took it well.

8.18 CRASHES IN CHICAGO

I recall a Chicago meeting with Fred Giles. We always took the cheapest accommodation, and in this case it was the YMCA. As we were trying to go to sleep, we were awakened by the most intense, loud crashes outside the window. It turned out that the inhabitants got their kicks by dropping glass bottles onto the concrete floor of the quadrangle we faced. In the morning we looked out to see the floor covered with broken glass. Fred was an expert on Chicago, and we went to a Danish Smorgasbord which had a small marionette theater associated with it – a tiny replica of the Kongestheater in Copenhagen. They performed, "My fair lady". At the end, the puppet manipulators stood on the stage – they looked gigantic, because we had associated the puppets with real people.

8.19 FLORIDA STATE, AND THE DOG DRIVEN CAR.

Florida State University had acquired a High Voltage Tandem Accelerator just after I got to South Carolina, and I went down to Tallahassee several times in connection with this. Merle Tuve spoke at the opening. I knew a radioastronomer there, who had emigrated from England, bringing his car with him. He had a large dog (Dalmation, doberman?) which used to sit in the passenger seat. However, this was of course on the left, it being an English car. Since the steering wheel was set rather low, this looked for all the world as if the dog were driving the car, with my friend in the passenger seat. Several near-accidents occurred with oncoming traffic as a result of this.

8.20 YALE

In 1963, my boss Tony French decided to go to MIT. I felt I too should start looking around. The usual process was to fill in a form and go to the "slave market" at the New York meeting, (where job interviews occurred) which I did, and got a few offers as a result. These were not real "offers" but a request to visit the establishment having a position available. More often than not it did result in a genuine offer. It was still a time when jobs were easy to get. The one which attracted me most was a research fellowship at Yale, which was funded to design a "meson factory" – a machine built primarily to produce pi mesons by striking a heavy target with an intense beam of 600 MeV protons. This was a new concept, and our end of it was to design the target setup.

8.21 VERNON HUGHES and GEORGE WHEELER

Vernon Hughes was the titular head, being Yale department chair at that time. Being chair, he had a front office which looked like that of a CEO – mahogany desk, very tidy, with little showing, indicating how focused he was. However, directly behind this executive suite was the real office – papers everywhere, untidy, just like mine. This is where the real work was done. It reminds me of a German cartoon I have showing a guy hunting over a desk covered with papers and the text "He who is tidy is just too lazy to look". Vernon was very dedicated. We had work on the Nevis cyclotron, and I recall he would take his sleep going to and fro between Yale and Nevis, which was on the Hudson just north of New York City. His wife also worked with us. She was Jewish, and he was of Welsh extraction – a curious combination. I drove to a conference in Boston with her one time, and we had some interesting conversation – Vernon had invited a famous orthodox Jewish physicist to dinner, and she had cooked swordfish, which, she had just discovered, have no scales and cannot be eaten by orthodox Jews, so she had to quickly replace it.

Most of the organizing was done by George Wheeler (who ultimately became dean at the University of Tennessee in Knoxville). I was hired to assist Brooke Knowles in doing the calculations, and we all worked together, with Sho Ohnuma, (who was designing the accelerator itself) in an old sea captain's house on a hill just above the physics building. Very little had changed in this building since its construction, so it still had a "widow's walk" on the roof, from which one could view New Haven harbor. Sho was very Japanese, and did his work sitting cross legged on his desk, which I found a little unnerving. We frequently had guests from Brookhaven, who were consulting with us, such as Ernie Courant. They were a very interesting bunch of people.

8.22 BRANFORD

Renting a house in new Haven proved a problem – in the end we bought a house in Branford, a small community along Long Island Sound. It had a village square, and all the accoutrements of a typical New England village – plus the problems. Town residents such as us could go to the park on Long Island Sound – where one day I saw a skunk crossing the road. It leisurely walked across, and nobody – naturally wanted to do anything about it. The Sound water was very cold for swimming, but the rocky area surrounding the beach was quite delightful. Our immediate neighbors were nice, but the general tenor of the locals was aggressive. I recall one neighbor had run his car into the post box of another, who came out irate, quite naturally, saying, "What are you going to do about it?" The offender merely shrugged his shoulders and said, "So sue me". We liked Brooke Knowles' family, but we found everyone hibernated in the winter, so only those friends we had made in the summer persisted. The local social life was not very active. Many of the locals worked in New York. They would board a train nearby at 7 in the morning, read the paper and play bridge till they reached Grand Central Station, returning home the same way. They had a life on the train.

8.23 RUNNING ON THE COSMOTRON

In designing the linear proton accelerator (for which I had been brought to Yale, and which ultimately was built at Los Alamos) we required the pion production cross sections from different elements, some of which were not available, so we started doing experiments on the Cosmotron accelerator at Brookhaven National Laboratory on Long Island, employing the experimental equipment set up by the Rochester University group, one leader of which was Adrian Melissinos. He was a delightful person, certainly one of the most polite individuals I have met, which made dealing with him very easy.

8.24 WOODY GLENN

We worked with Woody Glenn, the chief engineer of the Cosmotron, and quite a character in himself. Amongst other things, he had crewed for ocean yacht racing, Newport to Bermuda. I recall one time he picked me up at the Jamaica train station in the snow. We drove out to Brookhaven, along the Long Island Expressway. However, on turning off, the car spun completely around twice, ending up stationary pointing in the same direction we were traveling. Woody immediately drove on quite nonchalantly, as though nothing had happened. Meanwhile, I was biting my nails. We stayed at the dorms in Brookhaven, which appeared to have been erected by the military during the first world war, though it could have been in the second. The uncomfortable bunks were reminiscent of boot camp, and the shower stalls were of galvanized iron, from which the galvanizing had long vanished, and only rust remained.

The Cosmotron was one of the first of the large accelerators, and as such had immense magnets to steer the beam in an approximate circle – strong focusing (which makes the magnets much much smaller) being long in the future, a "gleam in the eye" of the accelerator designers. Since it was concealed by massive concrete shielding, all we saw was a hole in the wall through which the proton beam emerged, and traveled through the air a considerable distance at about waist height.

8.25 BOING-BOING

Our "bending magnets", which directed the beam, and detector systems, did not have vacuum chambers for the beam after it came out of the machine. Hence it was quite possible to walk directly through the beam, which would pass through you at waist level. However, a pulse came only every two seconds, so if you were smart, you could cross the beam between pulses. In order to deal with this problem, there was a bell which rang as every pulse came through. What I never could work out was, did the "dong" come just before, just after, or exactly at the pulse? For days after I left the accelerator, I could hear the damned "dong" in my ear – it persisted in my brain long after I went away.

8.26 GOLD PLATED EQUIPMENT

One interesting aspect of the electronic equipment was that it was all gold plated. Since the electronics was made on site, and was very temperamental, the designers had gold plated the front of the modules which slid into the racks, to avoid contact problems. Hence, the front of the rack was all gold – very thin of course, but giving an exotic appearance I have never seen elsewhere.

8.27 AN OLD STUDENT – FORSYTH

I was at lunch one time at the Brookhaven lab when I ran into someone I recognized. He obviously also recognized me. So I said "We clearly know one another. Was it in Australia?" "No, never been there. How about Canada?" "No, never been there." It turned out we had been at school together. He was a year younger than me, and had lunch at the school, as did I. I was in charge, and sat at the head of one table, and he was one of the eight or ten boys sitting at this table. He was an engineer named Forsyth working on how to get power into New York city without the objectionable overhead power lines. They hoped to use superconducting cables. The big problem was getting the current into and out of the cables at the end. Many, many years later I met him at a get together my school had in New York City. His wife had died, and he had bought a yacht on which he was sailing hither and thither around the world. He was in the process of sailing to Antarctica at that point.

8.28 A SHORT ENGLISH VISIT AND THE BBC

I had to go to England to give a paper at a meeting at the University of Manchester. It was fascinating to return to my old stomping ground and see the changes – as a child everything was black with soot – and to see the town hall and buildings a beautiful sandy color after cleaning was a delight. I also gave an interview on the BBC. When a ten year old child (before my voice broke) I performed with a school choir at the BBC studios on the square in Piccadilly in Manchester. They were experimenting with announcers having regional accents, and Wilfred Pickles, a native speaker, read the news. I think they got rid of him later. I found the television studios were in exactly the same building, and what had been a sound studio was now television. One thing, before the interview, they wheeled in a trolley and we all had a sherry. This definitely relaxed me to talk about the recent integration of the University of South Carolina, which had gone on without a hitch. I think there had been a lot of preplanning, which had obviously been a good idea. I talked about this. Now it seems so far away – our university has changed so much, it is difficult to recall what it was like while segregated. Later, several friends of my parents were delighted to have seen me on TV after so many years abroad.

8.29 BACK TO COLUMBIA

The project to build the meson factory was a three way competition between Yale, Los Alamos National Laboratory, and the University of Southern California, a group led by Reggie Richardson. It eventuated that the accelerator should be built at Los Alamos. As time proved, this was the right decision, but it was reached more by adept politics than physical argument. Anyway, the Yale project was terminated after I had been there a year and a quarter. I had been told this might happen, but was left with several options. One was to stay on at Yale and work on the Nevis cyclotron, which was being revamped. I had an offer to go to Brookhaven and work with Sam (Lindenbaum) and Luke (Yuan). This was an attractive proposition, since they were doing interesting work in high energy physics. In the end I opted to go back to the University of South Carolina as a full professor. The combination of both teaching and doing research appealed to me – and reaching a full professorship at that age (35) was attractive. I wonder now about my judgement, but I have no cause to regret my decision, in that I managed to do all the things I wanted to – although I never earned much money – but neither did I get fired – and it was fun. There is no point in doing physics unless it is fun!

9 TELEVISION AND THE SPACE RADIATION EFFECTS LABORATORY

9.01 TELEVISION

I had always been interested in television, however, the possibility of actually being on television struck me as being most unlikely. Rudy Jones, a colleague of mine had a friend in our State Educational Television Studio. At this time, in the early sixties, the outfit was young and small, but growing rapidly. So he introduced me, and I got my own program. It has always been my contention one should grasp opportunity with both hands, because it likely will not come again. This, I am sure was a case in point – in some ways it was a daunting project, since I had no background whatsoever of producing a show – I had never been involved with live theater, except for a walk-on part carrying off the bodies in Hamlet at school, and I am sure this offer was a rare event, because the whole character of television has changed to large and expensive productions. As it is, I had more fun and excitement out of this than most physics experiments.

9.01.1 SCIENCE TALKS

My program, called "Science Talk" came on once a month, with a repeat. It was introduced by "Jupiter", from the music of Holst's Planets Suite, a suggestion of my producer-director, Dave Smalley, for whom I have the utmost respect. He was the lowest key director I ever had, which worked beautifully for me – and most of the other prima donnas he had to deal with too I feel sure. At that time it was all black and white – color was yet to come. That led to some interesting possibilities. I decided to put color on black and white television using "Benham's top". This is an interesting optical illusion of the nineteenth century. Originally it was a top, with a diametrical line across it, the disk on one side of the line black , the other white. Then, there were some black segments of circles centered on the top axis. On spinning the top, the color sensors in the eye responded, (or recovered) at different speeds for different colors, with the result one saw rather muted colors. To my knowledge, this had not been tried on television previously – anyway, I made a disk and spun it in front of the camera – and sure enough one saw colors. Suddenly, all the cameramen vanished – I couldn't work out where. Turned out, they had bet their friends they could put color on black and white television – so they all came to see it.

The set was illuminated by Klieg lights – powerful filament lamps. This was because the cameras were not as sensitive as those today. The set was freezing cold as you started, but with the hot lights beaming down on you, it wasn't long before you started to sweat profusely. This is what lost Nixon the election about that time.

My shows were taped, but my director made me do the whole half hour at one shot – he said continuity is better. I did not really believe this but he was right. When I tried splitting it up, it proved a disaster. Mark you, we did do a lot of sequences too, because they had to be shot elsewhere. One time we went down to Cape Hatteras to do a segment on a reenactment of the Wright Brothers first flight. They actually flew at Kill Devil Hills some miles away, but that has become too built up to be useful. I had just learned to fly, so I piloted a Cessna 172 flying down. Landing at Billy Mitchell field, however, proved hazardous. There was a strong headwind on landing (that was why the Wright brothers picked this spot), so our ground speed as we came in was low. Suddenly, as we dropped below the sand dunes the wind stopped – and so did the plane. Anyway, we made a good if somewhat bumpy landing, shot the sequence with the simulated Wright bros plane, and started back. There was a humongous headwind again, slowing our little plane down to a groundspeed of 40 mph or so. Under these circumstances, it is a good idea to fly low, where the headwind is less – but it can be (and was) extremely bumpy. It was, without doubt, the most frightening flight I ever made, and Dave Smalley still blanches when the subject comes up. But we made it.

On another occasion we did a program on radioastronomy, and flew to Greenbank West Virginia in Coke Darden's Lockheed Electra. The cameraman had never flown before, and it proved most difficult to find the small Greenbank airstrip. West Virginia consists of parallel rows of mountains and valleys. We would fly up one valley, and down the next without seeing the strip. Just as the cameraman was going hairless, we saw the strip near the large antennas, and landed. We wondered whether flying over the big dishes would affect their results but they said not.

Many of the shows involved several topics, and I had to interview the speaker if there was one. Most speakers run on too long, and one has to develop techniques to interrupt them without it being too obvious. I would interrupt by saying, "I see," and then go on – but this did not work too well because the repeated, "I see," stuck out like a sore thumb. New topics and sports did well. I did one on ball games, and we showed how the Bernoulli effect played a part in golf, tennis, ping pong, baseball, basketball, and football. I well recall going onto the baseball field and having balls pitched at the camera. I also went onto the basketball court as they were playing. I felt like a midget.

9.01.2 SEVEN THIRTY

When they started up a new program called "seven thirty" I was invited to be a reporter. I was frequently teamed with Phyllis Giese, the wife of a famous football coach and it worked out very well. Since it was live, there were some interesting things happened. They say there is nothing that gives you ulcers like being on live TV. For example, I had interviewed Paul Dirac, the physics Nobel Prize winner. It was on 16 mm film, which was put on the "chain", so that I said live, "Today I talked to Prof. Dirac about how he came to be interested in physics". They were then supposed to run the film – instead of which, the black comedian Dick Gregory appeared, from another segment of the show. The chain was in the wrong order, but there was no way to change it, so the camera came right back to me open-mouthed. The ability to recover rapidly from a situation like this is the mark of a good TV personality, so I said, "Well, he told me…,etc., etc." On another occasion, I got a phone call at home that one of the interviewers was absent, and would I replace him? I said, "What is the interview?", and was told, "Oh, some science thing". At the studio it turned out it was an interview with the director of the state's VD program – and I know nothing about the scientific side of VD – but it was too late for me to back out. Phyllis Giese was also an interviewer – and the program was live. There was much trepidation about what questions would be phoned in. It turned out, they were all very meaningful – but I was very glad when that program ended.

9.02 MONTY PYTHON

My most interesting assignment came in the eighties, when I was called on to introduce the Monty Python show. Our ETV had an arrangement to show this English comedy, and they were looking for an English voice to introduce it. Originally this was to have been someone else, namely the secretary to the university, but I was lucky because he could not do it. They ran two sections of approximately half an hour, back to back. Each piece ran about five minutes short, so I had a five minute introduction, and a five minute conclusion. A delightful set was built for this, with a big coat of arms

(Cambridge), and a genuine silver plastic suit of armor. A picture frame was built over a hole in the wall. One of the production staff would put on a cavalier's outfit, with sword and feathered hat and stand in the picture. At the right moment he would reach out and tap me on the shoulder saying, "No, no – that's all wrong!" – the picture came to life! On one occasion, the idea arose for me to take the part of a char lady – a very common event on English TV. So I put on a dress and wig, and came on with a bucket and mop, saying, "Can I do you now sir?" like Tommy Handley in WW2's British ITMA (Its that man again – the war stopped for ITMA). Anyway, I thought the piece went very well, until my sons and I were watching it. "Good grief" said one, "Father's on Television in drag!"

9.03.1 SPACE RADIATION EFFECTS LABORATORY

Financial considerations dictated that the Cosmotron should be closed. It was mothballed, and I believe you can still see the old magnets. We were determined to continue this line of research, and a relatively new accelerator at Newport News in Virginia seemed perfect for us. It had been built ostensibly to study the effects of the radiation in space on different objects. However, so far as we were concerned. it was the ideal machine for what was then called "intermediate energy nuclear physics". The machine produced 600 Mev protons. It was a synchrocyclotron, and the identical twin of the machine at CERN the European Nuclear Facility, which I later visited. I have a vague feeling one accelerated clockwise, and the other anticlockwise. The Space Radiation Effects Laboratory, SREL as it was called, (now the Jefferson Lab) had several accelerators on the site, for checking parts for the space program, shuttles and such, against radiation damage such as occurs in space.

I have never been any good at raising funds, so we had no extensive grants. However, we got considerable free running time, and managed to cadge equipment from a variety of sources.

9.03.2 THE LABORATORY.

SREL is situated on the marshy peninsular running down to Newport News and Norfolk. We used to stay at a motel nearby, which had a rather crummy restaurant. One time we came back for our next run to find the restaurant totally changed, with nice decor and a good menu. It turned out that in between our visits the state of Virginia had converted from being totally dry to allowing alcohol. The result was that the restaurant had upgraded with the added profits.

One problem was that, also between runs, the staff would completely revamp the software for the IBM computers we used. This was a trial, because it took a day or two to get our own software to run again.

9.03.3 RUNNING AT SREL

We were normally allowed a few days on the accelerator, so we had to run shifts. I did not like the late night shift, which others preferred, so they would take over then. The big problem was relaying the information collected during a run at the shift change. Each bunch of experimenters explained what had happened to their successors. Invariably something was forgotten, causing difficulties.

9.03.4 FAST REPAIRS

On one occasion, the accelerator broke down in the middle of a run. There was something wrong with one of the copper Ds in the machine which were highly radioactive. We had to go inside the accelerator to fix it. We calculated we could only stay inside for ten minutes before we accumulated our yearly dose of radiation. So each of us went into the machine, and started fixing it. After ten minutes out we came and someone else went in. We did fix the machine!

9.04 JOE'S STONE CRAB RESTAURANT

As I grow older, I think more and more of an episode that occurred on a trip to a physics meeting in Miami. Arriving at the hotel, I met up with Willy Haeberli, a colleague from the university of Wisconsin, and we decided to go out to dinner. However, we had no idea where to go, so we took a bus and asked the bus driver. He recommended Joe's Stone Crab restaurant (a very good choice as it turned out) and told us how to get there by bus. You bash the crab legs with a hammer provided by the restaurant in order to access the meat. We returned by bus, and as we got off, encountered an elderly gentleman sobbing his heart out. It seemed that he was completely lost, and wanted to get home. He had come out of his nearby house and wandered off. We suggested that we walk around, and see if he could recognize his house. Sure enough, about fifteen minutes later this happened. I have never seen such relief and happiness suddenly suffuse someone's face. It made me realize, if I got Altzheimer's disease, this might be me in a few years. There, but for the grace of God, go I.

9.05 SAVANNAH RIVER PLANT

The availability of the Savannah River plant reactors not far away was an attractive proposition for experiments requiring a high neutron flux.

Rudy Jones and I worked with J.A.Stone and W.C. Pillinger for a while. However, the most amusing event occurred when it was decided to allow foreigners into the plant which, until then, had had the utmost secrecy. Since Tony French, our department chair, and I were both English at the time, it was decided to make us the Guinea Pigs – we would tour the plant to make sure there were no glitches, should the Russians come. All went well, and we had as nice visit. Actually, we had seen some things we should not have, but we didn't know it!

10 FLYING

As a child, I had always been interested in flying. I recall in the mid 1930's going to an open field between Bolton and Horwich in England, where a flying exhibition (biplanes I think) was being put on by a barnstorming outfit. However, my first flight was between Sydney and Canberra, as I mentioned before. I will never forget what a marvelous plane I thought that was. We traveled rapidly to Canberra, compared with road transport. I had always believed I could not get a pilot's license, because I am colorblind. That would certainly have precluded being a pilot in the British Air Force, or I might well have joined up, because I had always had such a desire to fly a plane myself. Imagine my surprise, then, when I found one of my students, whom I knew to be colorblind, was a member of a local flying club, which had just been set up. He told me that colorblind folks could hold a pilot's license, with the restriction they could not fly at night. I joined immediately!

My interest in flying had been piqued by flying with my colleague Fred Giles, and also by going to and fro across Long Island Sound when I was working at Brookhaven National Lab. We had a contract with a New Haven firm which flew us the thirty miles or so from the small New Haven airport, to the even smaller one at Upton, (near Islip) on Long Island. The alternative was 150 miles or more (plus New York traffic) by road. If I was the only person going, the pilot would let me take the controls, and I recall the first time he let me turn and make the approach, and finally, just as we were about to land, I said "Here, perhaps you had better take over". Actually, I thought I was doing rather well – but had no idea how to land! On another occasion, we flew in fine weather over the sound, but the Brookhaven airport was fogged in, so we had to fly back. We went to and fro three times during the day, but the coastal weather did not clear, so we had to give up.

On yet another occasion, it had snowed heavily in the night, but the plows on the Upton airfield had not made much headway. As we came in I could see one plow had made a long snaky line on the runway. Making our approach, the plows took off in all directions. The plane landed on the S shaped runway with no problems.

I should mention there was one other way between New Haven and Brookhaven – the concrete ferry that plied between Bridgeport, and Port Jefferson. It was a very old car ferry, and one time I went down into the hull, and found it had rusted through here and there. At first they had put on one or two patches of concrete over the holes, but as time went on I found they had put concrete patches on concrete patches. Nevertheless, it did not sink.

10.02 FLYING TO SREL

Coke (Colgate) Darden joined our collaboration, first to run at Brookhaven on the Cosmotron, and then at the Space Radiation Laboratory (SREL) at Newport News in Virginia. Coke had a number of pre-WW2 planes including the Lockheed Electra (Amelia Ehrhard got lost in such a plane). When told we had running time on the cyclotron at SREL, we would load the plane, climb aboard, fly to Patrick Henry field, and be running in a few hours. I flew the plane from the copilot's seat occasionally. The first time I flew, after I had been flying the plane awhile, an alarm went off vigorously, causing extreme anxiety on my part. Coke said, "Oh I forgot to switch to the auxiliary tanks!" Then, in midwinter, the plane iced up on the ground in Virginia. The instruction books say you should not use hot water to de-ice the plane, but employ antifreeze. We only had a little antifreeze, and it didn't do much good, so we tried hot water, which worked much better. We concluded that the instruction book was written for someone with temperatures well below zero – not us, just a little below. Anyway, as it was, we took off with small icicles hanging from the fuselage, but by the time we landed in Columbia, they had all gone.

You could not see the runway ahead from the cockpit in this tail-dragger plane of an earlier vintage, so you either steered by looking out the side windows at the edge of the taxiway, or steered an S or snake like path down the tarmac, again looking out the side windows.

10.03 LEARNING TO FLY

I was taught at a flight school run by a couple of ladies with flying in their blood. I had a number of instructors. For the most part, these were young men who wished to increase their flying hours to improve their chances of being hired by an airline company. However, there was one very old man (by my standards then – he was at least 65!). He was short, and always carried a special cushion to sit on – which I did too – it proved invaluable, allowing one to see over the nose of the aircraft. I recall on my first cross country flight, I took off from Columbia, and he promptly went to sleep, until we reached Charlotte. On my attempt to get a license, one of the ladies (Frances) was the examiner, and the only thing I did wrong was forget to take the flaps up after an aborted emergency landing.

10.03A MOONSHINE AND THE CESSNA

Whilst I was being taught, we would fly over the Congaree swamps below Columbia. One very calm day, we were flying across a big dark swamp, when I noticed a plume of smoke rising straight up in the distance, a couple of miles away. As we got nearer, the plume suddenly quit, and as we flew over, all you could see was a dense jungle. Once we were a couple of miles beyond, the plume mysteriously arose again. My teacher said, "It's those damned moonshiners". The revenuers (Revenue Officers) would try to locate illicit stills by flying over this area in a light plane to spot the position of the still. The moonshiners, knowing this, would rapidly conceal the still and douse the fire as soon as they heard the plane.

10.04 THE BREAKFAST CLUB

I belonged to the "Breakfast Club". Once a month we would agree to fly in to some airfield on a Sunday morning. The landing of ten or more planes at the same airport, all within half an hour or so was quite an event. The landing was observed by the early arrivals, and the individual making the worst landing received the award of the "bouncing ball". I never actually received this, but I was involved one time. I had made my approach and observed the wind sock to ensure I was flying in the right direction. As I descended I noticed a plane approaching from the opposite side. "That idiot will soon see he is flying the wrong way," I thought – but he did not! Pretty soon we were committed – we had to make a landing come what may, so we landed on the same runway from opposite ends. Luckily, it was a long runway, so we both stopped before the taxiway, which was in the middle. He got the bouncing ball!

I had several interesting experiences while flying. For example, I had landed at Chester airport , on one of the triangular tarmac runway systems the South Carolina counties erected before WW2. The field was covered with grass which had grown way higher than the cockpit windows. Still, I knew from the windsock again that I was on the right runway, and proceeded to take off. Imagine my horror when the plane rose above the grass, to see a crop duster taking off on the runway sixty degrees away from me, so we were on a collision course. We both bore off away from one another, and nothing happened. It seems crop dusters pay no attention to the wind. The tanks of insecticide were on the other runway to mine, so after filling up, he just took off. They say there are old pilots, and bold pilots, but no old, bold pilots!

10.05 SEMI-SKYDIVING

Flying can take many forms, and I have been interested in all of them. I have never done skydiving, but there was in Pigeon Forge Tennessee, a very ingenious device, where an enormous electric motor below the wire mesh floor, drove a propeller taken off a large plane, to provide a vertical upward draught of air. The force of this draught kept you floating in the air if you jumped over it. You wore a clown suit, to provide more drag, and were assisted by an individual in swimming trunks, whose air drag was so small he did not float. After a little training, you could move about over the fan to and fro and if you fell off there were cushions to stop your fall. The fan speed could be adjusted to make you go up or down. That is the nearest I got to skydiving!

10.06 THE BLIMP

Whilst at a conference in Miami, I noticed an advertisement that for $5 you could go for a flight in the Goodyear blimp, Mayflower. I went down to Macarthur Causeway, but it was a very windy day, and they were unsure whether to run the dirigible. Ultimately they got the go ahead, and I was lucky enough to sit next to the pilot. Compared to a conventional aircraft, the controls on the blimp were very primitive. As I recall, the rudder was controlled by foot pedals, and the elevator by a large vertical wheel which sat on the right side of the pilot. The lift was provided by ballonets containing helium inside the hull, and the hull itself was kept rigid by a force of air provided by the propellers, which directed some of the air into a chute going into the hull. The gondola hung by wires from a strong support on top of the ballonets. Strings were attached to various openings to control the shape and other items. They had wooden handles you pulled. We took off smoothly, but there was clearly a strong wind aloft, and we moved slowly towards the coast. One interesting item, a sudden side blast of wind caused the blimp to roll about the center of the hull, so the cabin swung slowly from side to side about a bow to stern axis by a small, but noticeable amount. Having reached the coast, we went along it a short way, then returned to the landing field. The combined wind and speed of the blimp meant we got back in no time at all – probably traveling in excess of sixty miles an hour. We were flying at fairly low altitude – less than a 1000 feet or so. Our descent was most interesting. The pilot adjusted the speed to just buck the windspeed. Then he set the elevator to come down. Hence we came down almost vertically, just moving forward very slowly. The ropes hanging off the blimp were caught by the ground crew, anchored down, and we were able to get off.

10.07 PARASAILING

I had noticed these colorful parasailers over Miami and the keys. Very attractive – but it looked dangerous – nevertheless, I took a trip off an inlet near Key Largo. The speed boat took off at high speed into the wind, the parachute was released behind the boat, and I climbed into a harness – essentially a seat attached to the parachute. Then I was slowly reeled out to an altitude of about a thousand feet. You got an excellent view of all the little islands in the vicinity which I did not even know existed, since they were invisible from the road. As I was reeled out, the cable periodically emitted an immense "sproing" sound, terrifying me. It turned out the way the cable was wound, it occasionally jumped from one winding to the next, giving this sound – quite a normal thing! After we had tooled around a while, I was reeled in. The speed boat driver then adjusted the speed so that I was traveling along quite fast with just my feet and legs in the water – it was quite a pleasant experience.

10.08 HANG GLIDING

I had always wanted to hang glide, but the thought of running down a hill and jumping off a cliff did not appeal to me – particularly as I reached the age of 65. However, at SWIM (the Unitarian Universalist Southeastern Winter Institute in Miami) one time, they had organized a hang gliding expedition off Biscayne Bay. This sounded much more in my line, since it did not involve running or jumping. We went down to the dock, near the old Pan Am terminal in Cocoa Beach as I recall, and our pilot was putting together a very large hang glider – big enough to hold two people. This was mounted on the stern of a power boat, and we drove off. When it came to my turn, I lay down on the support parallel and close to the pilot. The boat applied power to go fast into the wind, and we were reeled out on a long steel wire, and went up like a kite to altitude – somewhere between one and two thousand feet. The pilot released the wire, and we started to descend, gliding around close to the shore to see the tall buildings in the city center of Miami. In the mountains, of course, you can use the updrafts to stay aloft, but there are generally no such things at sea. However, the pilot said, if you were a few hundred feet above the tall buildings in the city of Miami, there was an updraft provided by the wind striking the front surface of these buildings, and you could glide to and fro and stay up. There was a problem however. Inhabitants of these buildings would sunbathe in the nude on the roof, and they became irate at this glider going up and down above them with the pilot looking straight down!

The pilot kept asking me if I was alright – which I was, and enjoying it thoroughly. It turned out that the last passenger had thrown up in mid-air, and the pilot was afraid I might do the same. To land, we glided near our boat, and came down on the water – the glider had floats like a float plane. The landing was very gentle, a rope was attached to the nose of glider, and we were reeled onto the boat, to repeat for another glide. It so happened I had arranged to get married that evening. Somehow, this was mentioned to the pilot who said, "Hmm – I wonder which is more dangerous – getting married or hang gliding?!"

10.09 KEN PURSER

One of my former students, Ken Purser, had bought a Bonanza airplane, and visited us. On landing, he taxied to the fixed base, and parked the plane, leaving the engine running for about ten minutes, which kept us all waiting, since we could not approach the plane. Then he stopped the engine and disembarked. It turned out that his wife had a baby, and to make sure its ears did not get plugged, she was feeding it during the descent. On landing, the baby did not want to stop, so Ken let the engine run until feeding ceased! I flew the Bonanza, which seemed to have an amazing predilection to fishtail, and if you tried to compensate you made it worse!

10.11 COKE'S PLANES

Colgate W. Darden had accumulated a number of antique aircraft – all pre-world war two, but all flyable. One in particular, the Dolphin, had originally been built by Mr. Douglas – for Mr. Boeing, as I recall. This was a twin engine amphibious flying boat, with the engines sitting above the high plywood wing, and having a small wing between them. It was used a lot by the navy in world war two. I flew it only off land, and it was rather slow to respond to the controls – to go into a turn you had to hold the wheel over for quite a while. Landing on water, you pumped a large lever between the seats up and down, rather like the hydraulic jack used with cars, to retract the wheels. I never did land on water, but Coke tried this once, and said the plane filled up with water on touching down, because the sealing around the windows was shot. Coke also had a DC2 – pre-curser to the DC3 or C45, a Swallow, and the first corporate plane – a Spartan Executive. The pilot's seat on this plane had a habit of coming adrift, so it would slide to and fro. This was a significant disadvantage on take-off, because, as the plane tilted back, to stop it sliding, the natural inclination was to pull back on the wheel, which had the disastrous effect of making the plane tilt steeply. To avoid this, Coke, chained the seat firmly to the floor.

11 LOS ALOMOS AND LATER

11.01 LOS ALAMOS

As I mentioned, the linear accelerator (pion factory) which we had designed at Yale, was ultimately built at Los Alamos in New Mexico, primarily for political reasons, as is so often the case, but in fact it proved a good idea in many respects. I went out there to a conference as it was being built, and saw the hole cut for it in the "tufe" – soft volcanic rock which can be sliced with a bulldozer. Dr. John Warren from Vancouver (with whom I worked in Australia) also attended. His wife had had twins at the same time my wife also gave birth. The doctor had not realized there were two babies, so Mrs. Warren was taken back to the ward before they realized a second one was coming!

11.02 AN ACCIDENT ON THE LOS ALAMOS ROAD

We would rent a low cost auto in Albuquerque, and drive to Los Alamos. One time, I rented a Volkswagen Beetle in Albuquerque on a cold wet December day. As I commenced to drive to Los Alamos, it started to snow. By the time I had driven through Santa Fe, the snow was getting deeper, of course, because the road up to Los Alamos is at a considerably higher altitude. I was driving at a reasonable pace – maybe 30 mph, on a long straight road, when suddenly I noticed I was going sideways. The steering wheel had no effect, nor the brakes. Under such circumstances, one's whole life is supposed to flash before one, but in my case I thought, "What the hell shall I do?"

In fact there was nothing I could do, when I noticed ahead a turn in the road. Now this mountain road had regions of flatness, punctuated by driving beside a precipice. The question was, at the turn, was this a flat place or a precipice? It was very dark at the time, so I could not see ahead. The car went through a barbed wire fence into a flat field, then through another barbed

156

wire fence into another flat field, after which the car stopped, and I offered up a prayer. Suddenly, a guy appeared. He had seen the event, and came to help. It turned out, although the car was badly scratched by the barbed wire, the mechanism was fine. He held the barbed wire up while I drove under it back onto the road and continued on my way, driving around Los Alamos in this beat up car for the following week or two. The following Saturday, I went to a party, and a guy there said, "I bet you don't know who I am?" I said he was right. Then he said "I'm the guy who helped you out on the road last week", It was dark, and I was badly preoccupied at the time, so he was not offended – but he had been a big help stopping the car from getting even more badly scratched.

11.03 HOT SPRINGS

Since I had few friends at Los Alamos, one of my favorite pastimes was to go into the Jemez mountains nearby for long walks. The mountains form the rim of an immense caldera, (the vast crater of an extinct volcano) and there are hot springs at various points. For one hike you would leave the highway, and proceed to walk uphill and down dale for two or three miles, before reaching the hot pool. It was delightfully warm and relaxing – just like a hot tub or Jacuzzi. Most people arriving there had not bothered to bring swimming togs, so one just undressed and got in. The people who came were always most interesting, and there is something about being in a hot tub which leads to delightful conversation. Anyway someone complained to the park service about the skinny dipping, which put them in something of a quandary. In the end, they put up a sign saying Mondays, Wednesdays, and Fridays you could go skinny dipping, the rest of the time a swim suit was necessary. I have often wondered who, in the park service, hiked the six miles or so to see if anyone was naked, and if so, what they would do?

One individual I met there had walked a tightrope between the twin towers in New York – that is now a part of history. One time, I went into another pool, and discovered a most remarkable tickling sensation all over. I looked down to discover I was being eaten by guppies. These tiny fish were nibbling at my skin, thousands of them. They were probably getting me quite clean, but I rapidly emerged. I came to the conclusion that the pool environment was probably ideal for guppies. Some individual with a tank of guppies that they did not wish to pour down the sink, had taken them up there and released them, where they proceeded to breed!

11.04. FOLK DANCING

Los Alamos had one of the oldest international folk dance groups in the nation. This was partly because, when the Los Alamos laboratory was started in the early forties, the staff came from all over the world – and there was nothing to do except work – radio did exist, but no television, etc. – so, the folk dancing started, in the log camp building originally, and it has gone on ever since. They have the most amazing collection of recordings. Then again, one of the Israeli scientists was an expert on Israeli dances and taught us – outside on the green. One of the advantages of certain Israeli dances is that you form a line, so if you don't really know it, you join the end of the line and fake it until you can really do it.

11.05 THE DOWNHILL SKI RACE

Los Alamos has excellent ski slopes within a few minutes of the lab. However, unlike other ski resorts which promote the advantages of their piste, Los Alamos does just the opposite, because the workers there don't want to crowd their slopes. Hence, looking at their brochure, you will see something like, "Of course, you must remember this area was used for testing explosives in the past and it is uncertain whether it was completely cleared". Anyway, it is a good place to ski, and my colleagues and I made good use of the snow in winter. We had one Swiss in our group, Jean Claude (or was it Jean Pierre?), and one year he brought with him equipment for downhill skiing – the numbers which you tie on your chest, the flag poles around which you ski, etc. We went out as a group one Saturday, and set it all up. We also brought a keg of beer. As the afternoon wore on, the keg got emptier and I got more mellow. The others had done the downhill race, but I had no intention of doing so, being a neophyte. However, encouragement from the others, together with the beer, finally got me to agree to go. I started off down the slope, which did not seem to be too bad to start with. What I did not know, because I could not see, was that the hill dropped away very rapidly after you had gone a short distance. It was by then too late to stop, so off I went, merrily rounding the poles. In fact I did very well. The reason was since I was nearly the last, the others had skied a groove all the way down. and so long as I stuck in the groove, I was OK. Of course, I fell over a few times, but rapidly picked myself up. Anyway, much to my surprise, I was not last, but somewhere in the middle of the participants, and was congratulated on my skill. That was the one and only time I was in a ski race.

Jean Pierre was also interested in climbing, and it became a habit to climb down the wall of the canyon, (on the plateau at the top of which the pion factory was situated), cross the valley, climb up the opposite wall and have lunch at the excellent New Mexican cuisine restaurant situated at the top, then reverse the procedure to get back.

11.06 LEARNING TO RIDE HORSES

Since my children were coming out to Los Alamos for the summer, I attempted to organize things for them to do. One possibility was to learn to ride. New Mexico is an ideal place for horses, and I had been told that down by the Rio Grande, a couple had set up a riding stable, employing Indian land which was essentially desert. I went down there, and the lady in charge was very helpful. "Have you ever ridden before?" she asked. "No," said I, "I have never been on a horse". "Would you like to try?" she asked. "Very much," said I. "Well, there's a horse outside," she said, and helped me mount it. "Now," she said, "if you pull the left rein, the horse will go left, and the right rein, it will go right. If you want the horse to go forward, jiggle the reins. If it won't go, kick it." With these instructions, I set off down the path. After we had been going a while, the horse stopped, and no amount of kicking would make it move. I bethought myself that perhaps the horse had seen a snake or something in the path previously, and was scared of that region. So I turned the horse, and we set off through the brush. After a while, I steered it back to the path, and all went well.

When the kids arrived, we went to the stable, and my wife was insistent that someone accompany us who knew how to ride. A young girl was available, so she went with us, my wife staying behind. After a while walking the horses, she asked whether we would like to try another gait, so we trotted awhile. As I recall, I did not enjoy trotting. It seemed to me that the back of the horse came up to meet me, just as I was coming down. I now know why they have jockey briefs. Then we cantered. This I liked, and I can understand why the pilgrims used this to get to Canterbury. Next the girl said, "How about a gallop?" I was suspicious of this, but the kids were excited. At first, one might think that riding along on a horse at such speed would be terrifying, but actually, it proved quite comfortable, and even pleasant and exhilarating. My wife would have gone hairless had we told her what we had done, but luckily, all went well. I recall one horse was huge and called, "Big Red". They also had mules in the stable, and I learned that mules like to work in pairs, as in pulling a stage coach, since many movies were made at this site – the "Old West!"

11.07 RAFTING DOWN THE RIO

One of the summer relaxations was rafting. The rafts were generally large and comfortable, although considerable excitement was generated when we went over rapids. The canyon on the Rio Grande was famous, and I never ran that. Once in the canyon above Taos, the walls were so steep that you could not get out for many miles, when it came up to ground level. However, there were many other rivers nearby, and I went a few times. Even in mid-summer, the water was freezing cold, since it came straight from the Rockies. We put in near where the artist Georgia O'Keefe lived, I recall. The individual running the trip had brought a couple of large rafts, but also one or two small inflatable kayaks. I rode in a large raft, but after a while I was asked if I would like to try a kayak. I got in, and for a while all went well. However, in running one rapid, I bumped a raft, and fell in. The water was running fast and I could not get out. Over the next rapid I went, in the water. Then two or three pretty large rapids. You are told always to go over the rapid feet first under these circumstances, and this I did, even though utterly terrified. After the third rapid, the water leveled off, and I managed to paddle to the side and get out. Amazingly, even though the water was so cold, I was not. The reason was that terror generates a large amount of adrenaline in your body, and as you burn this up, it keeps you warm. I changed clothes, and never even caught cold, much to my surprise.

11.08 HANS BETHE

One of the fathers of the atom bomb, and of nuclear physics in general, was Hans Bethe. One summer he had an office next to mine. He was a most friendly person, still with a strong German accent. He came to give a talk at the Citadel one time, and I mentioned this to Helen Rader, a close friend in the Unitarians in Columbia. "Oh", she said, "He used to teach me, at Cornell in 1936 when he was a junior instructor", so I took Helen down for his talk, and we sat at dinner with him and his wife. "When was it again?" his wife asked, "1936? That was before I met Hans – we got married and I came to the USA the following year". Unfortunately, Bethe died recently.

11.09 HERMAN FESHBACH

I had another interesting individual adjacent to me in my lab office. We got talking, and I mentioned we could not afford to send my son to MIT because there were no scholarships. The guy commiserated with me, and fully agreed. I later found it was Herman Feshbach, the department head at MIT at that time!

11.10 BERLIN BARREL ORGANS

I had a conference in Berlin about this time which coincided with a barrel organ festival. Apparently, people buy and refurbish old barrel organs, and get together to play them on street corners here and there. They wear garish outfits (the guy staying in my digs was dressed like an ambassador). The director of the conference had hired such a one, and I got to play the organ. However, before I could do this, I had to wear a fantastic hat covered with artificial flowers.

We took a water tour, and I had no idea how many rivers and canals the city of Berlin had. The Berlin wall followed us around – you could see it, then it would vanish, and start again somewhere else. At the Brandenburg gate there was a stand where you could peer over the wall – now long gone I am sure. I bought a Tee shirt saying "Bier hat dieser wundershoener korper gemacht" – mine was a very large body.

12 MUNICH

12.01 MUNICH

In 1972 I decided I needed to take a sabbatical, and work on my ion – solids interaction experiments. Two options appealed – Aarhus in Denmark, and the Ludwig Maximilian's (mad king Ludwig's father I believe) University in Munich. Aarhus would not supplement my sabbatical income, whereas Munich would – and with a family, that proved very necessary. So Munich it was, and I arrived there about eight hours late because of aircraft problems. Nevertheless, I was met by two or three department members. I felt like a piece of seaweed after the long flight. My mentor, Lehrstuhl Professor Sizmann had arranged I should stay at a very nice pension with Frau Von Beckedorf, opposite the Kunstakademie on Akademiestrasse, and near the Siegestor. My office was adjacent to the Amalienstrasse gate of the University. The room, occupied by several students, had large windows. Herr Bell was given the job of helping me, but Carl Rau was the experimentalist I worked with, and Constantine Varelas the theorist. He was Greek. I don't speak Greek. He didn't speak English, so we were forced to speak in German, which proved a great boon – I could understand him far better than my German colleagues, because having learned German as a foreigner, he used simple constructions which I could understand. Only once did we have a problem. He invited my wife and I to dinner on "Sonnabend". This I assumed was Sunday evening. It turns out this is an expression used in North Germany, but never in Bavaria, for Saturday night. We did not appear Saturday night, but Sunday night, when the problem was cleared up, accompanied by considerable hilarity.

12.01a AMALIENSTRASSE

The University gate leading to the Amalienstrasse was a severe temptation.

Directly opposite the gate was a Konditorei – a pastry-shop having the most amazing cakes, which they sold by the slice. My favourite was the Prinz Regententorte, which was unbelievably delicious, and available nowhere else in the world. They also sold marzipan cookies, shaped into the most amazing forms, fruit, furniture, you name it. No wonder I gained weight! Further along was the Stop-In Pizza. This Pizzeria with a most American name was run by Italians, who served the pizza on wooden plates. After the meal, the waiter calculated the cost in Italian, "Uno, duo," etc., then translated to German, "Vier mark funfzig." Beyond this cafe was a hole in the wall which sold delicious crepes. It had many bottles of schnapps which you could sprinkle on the crepe. The open selling of liquor, particularly at street fairs and Fasching and such would have driven a Southern Fundamentalist Baptist wild.

12.01b LUNCH AND BEER

In common with other Europeans, Germans like to take their time over lunch. As a Beamter (state employee) I received tickets I could use for lunch, but only for certain restaurants. So I would go out with other faculty, and it was an enlightening experience. One of my colleagues was rather contentious, and whatever he ordered, he would always complain. The bedienung (waitress) would answer back, so a vigorous exchange would ensue. I was most embarrassed, until I discovered that the waitress expected this, and would have been disappointed had it not occurred. It showed the individual took a deep interest in the food. I also learned to have a beer with lunch. In America, if I had a beer or wine with lunch I would go to sleep in the afternoon – but if I did not have a beer in Germany, people would ask, was I sick? I found I could consume 200 cc of beer without ill effects – Munich beer is relatively weak, and ultimately it became a habit, a pleasant one to go with the Schweinshaxen, quark and other delicacies.

Beer was the only thing delivered to our apartment, and played a large part in the lives of all Bavarians. Members of the physics department went on a trip to Kloster Andechs on one of the Sees (lakes). We took a boat up the lake to a landing, then had a long uphill walk to the monastery which was famous for its beer. A vast number of people were drinking there, and I must confess, the monks really know how to brew. During Lent, they are allowed to brew a stronger beer, "Starkbier," as it is called. The tale here is that because of fasting, they might not be allowed to drink the Starkbier during Lent. So they asked the Holy Father if it was OK. He took one sip and said, "Ugh – awful – yes, you can drink that."

12.02 SKIING

The Bavarian mountains are excellent places to ski, and I frequently made trips to Seefeld and many other resorts. A number of curious events occurred in the mountains.

12.03 LOST SON

In the spring I took my eldest son Chris skiing on the Zugspitze, Germany's tallest mountain. From Germany you go up the mountain on the cog railway (Zahnradbahn) which ends its journey inside the mountain. There is also a passage way to Austria, which owns the other side of the mountain – so you had to show your passport to get by (not any longer). Skis and such could be rented inside the mountain. This Christopher and I did. You can ski down from the rental shop on a pleasant smooth surface, but this is edged by precipices. We skied down once or twice very nicely returning via the chair lift. On the next turn down, I got ahead of Chris by about a hundred yards. With no warning a snowstorm descended on us with great vigor, and visibility suddenly decreased to zero. Since all one had to do was to continue on downhill, this did not seem to be so bad. There was no way I could go back, so I continued to the bottom and waited – and waited, and waited – still no Christopher, by which time I was going hairless, trying to see through the falling snow, asking people who had just come down if they had seen him, all to no avail. I had visions of Chris going over an invisible precipice. At long last, Chris appeared. It turned out he had fallen in the snow, and had considerable difficulty getting back up – ultimately, some other skiers helped him, and he descended – unhurt. We called it a day after that.

12.04 THE SWINGING CABLE CAR

On another occasion I took my other son Michael up the Zugspitze on a very rough day. From the end of the cog railway, a cable car takes you to the Gipfel – the crest. We got on the cable car to come back, to find it was swinging violently too and fro across its direction of travel – very disturbing. I was afraid the car would hit the pillar supporting the cable. Suddenly the cab stopped. Shortly it began to move rapidly – and I realized the operator was watching. and arranged to move the car when it was swinging in the opposite direction – away from the pylon – a dicey operation, and one requiring a good deal of skill, but it worked, and we arrived back safely

12.05 PASSPORT PROBLEMS

On one occasion, I took a bus trip to go skiing at Seefeld in the Tyrol. It was a lovely day, but after we had been going a half hour or so, I found I had left my passport back home. "No problem," said the other bus passengers, "The Zollamt (passport officer) will let the bus through as a whole, since we are all together". This happened, and I spent a nice day skiing. We had been told to meet at the "Treffpunkt" (meeting point) at a certain time in the afternoon. I got there – but no bus. It dawned on me I had misinterpreted the leader's expression of time. This was always a problem in Germany, for example, half past six can be "Halb sieben" (half seven) or "Dreizig minuten nach sechs" (30 minutes past six).

Anyway, I had got there an hour late. What to do? Take the train. So I marched down to the train station which was nearby, and booked for Munich. I got on the train with great trepidation. Would the Zollamt put me in jail at the border? etc., etc.

As we approached the border I got more anxious, and asked the two other passengers in the compartment where we were. It turned out they were genuine Tyroleans, and I couldn't understand a word they said, even though it was in German. We went on, and finally I recognized one of the station names, and it was Bavarian. Terrific sigh of relief. It was a Saturday night, and the Zollamts were probably ensconced in a warm pub somewhere, instead of being on a cold train.

12.06 DACHAU

The horrors of the German death camps had impressed me ever since they were news in world war II. I found, however, my kids knew little about them. So it was I took Christopher to Dachau, which was a short bus ride from Munich. Although the site and the surrounding barbed wire fence were still there, most of the cabins had been torn down. Some had been left as examples of the miserable conditions, and the cremation ovens remained. They gave me a sinking feeling just to look at them.

As we were walking around, we saw a small group of people talking to a catholic priest. We went over and listened. It turned out he had been imprisoned here. Most people thought of the Jews when concentration camps were mentioned, but he pointed out that Catholics, gypsies, and other unwanted individuals were also held. Once a year he came to hold a service for the inmates who had died, and this was the day. His description of life at the camp was graphic – they were forced to work pulling carts and do other menial tasks, with virtually no food or attention – so the dead had to be removed each day from the cabins. Many died because they just lost hope. However, we asked him what was the most important characteristic which kept people alive. I shall never forget his reply. It astounded me. He said – a sense of humor. He said if you could make fun of the most appalling conditions in the camp you might survive.

Perhaps, when things get bad, a sense of humor does keep you going. When I was in Russia, before the perestroika changes, I found that making jokes about the administration was one of the facts of life. Of course, although bad, the situation was nowhere near anything like Dachau. Nevertheless, this was one important thing which helped Russians survive.

13 TRAVELS

13.01 SABBATICAL IN BRIGHTON AND SOUTH AFRICA

In 1982, time had come for me to take another sabbatical. I had been many years out of England, and wanted to see if I could still live there – so I decided to take a sabbatical in the south of England, where ostensibly the weather is warmer.

Brighton is where the Prince Regent kept his mistress, Mrs. Fitzherbert, in the early nineteenth century, and the "Pavilion", the bizarre palace built on designs of what the architect thought was a Byzantine palace, is still there. It formed the basis for the design of innumerable funfairs ever since.

13.02 MISERY IN HOVE

I arrived in Brighton early in the new year of 1982. It was bitterly cold and miserable, and continued that way until after Easter. I did not see the sun but once in that period, and the dull and rainy weather made me exceedingly depressed – I had not realized the bright sun of South Carolina has a distinct effect on your temperament. Furthermore, finding accommodation proved very difficult, and the guy I had come to work with had absconded and vanished just before Christmas. I never did find out why – I think he just got disgusted with physics. I had thought to take a small cozy apartment, but I ended up with a cold, draughty, high ceilinged top floor flat in Hove. It was two blocks from the sea, and was probably what the Victorians regarded as the ideal summer accommodation.

I spent the first night on the floor in the kitchen, on a blanket, with the oven and stove going full blast. The gas fire had little effect on the living room (in typical British fashion, it was fed by florins, half crowns, or shillings or something to keep the flame going, extinguishing being presaged by the gas flame slowly dying, when one dashed over and fed another coin into the meter slot). I don't believe during the whole time I was there that the flat ever got really warm. Furthermore, with the absence of my colleague, I really got very little work done. Two things made up for this miserable state of affairs, I got in some excellent church bell ringing, ringing the changes, and went to some delightful barn dances. Sussex proved an excellent ringing venue. Apart from one or two peals I rang with the congenial crew of ringers, there were plenty of practice nights, and we rang at two churches every Sunday morning, stopping off at the railway station for coffee on the walk between them.

My work was at Falmer, the University of Sussex, whose vice-chancellor (president in America) was my old thesis supervisor, Denys Wilkinson. Denys had a rather messy divorce, in spite of which he got a knighthood. Shows how things have changed in England. I only saw Denys one time at Sussex. The university had been built in a new, red-brick design and did not impress me. There were cows in the fields outside the sylvan setting, but to get to the university you had a long bus or train ride from Brighton. The double decker was always full of coughing, sneezing people, and I had a perpetual cold, which did not improve my temperament.

13.03 THE MOTOR CYCLE

As a result, I decided to get a bike, and ride to and fro. I saw an advert in the paper, and went to see the bike. It turned out it was a motor bike, but the wife of the owner was anxious he got rid of it – so it was a good deal, and I took it. The vendor threw in a crash helmet, and large leather gloves. It was a Suzi – i.e. Suzuki, and ideal for my purpose. It was only 50cc, but a real two stroke motorbike and it behaved better than my previous 125 cc. BSA bantam. No longer did I have to carry a spare plug with me, in case the original oiled up. It carried me to and fro to work, and once I started using it I recovered from my colds. I used it to attend various barn dances, and when my second wife, Gertrude, came over, I borrowed a crash helmet for her, and she came too, riding pillion.

13.04 MY FIRST MOTORBIKE

My first bike, the BSA bantam, was purchased in Bolton, England for about 50 pounds. I remember, because I paid with a treasury note. These were large, unlike ten pound notes, and simply printed in black ink on one side. They were therefore easy to forge. As a result it was common practice to ask the guy who was paying you, to sign it on the back, which I did, following after many others. I used this bike to commute between my home in Bolton and Cambridge for many years. I used to stop off at my Aunt's in Kirby Muxloe. The Bantam, as its name suggests, was quite small, so I remember one occasion, traveling through the peak district in Derbyshire when all the 18 wheelers were stuck in the snow. I simply carried and pushed my bike through the snow drifts and came out on the other side – where there were 18 wheelers going the other way. To keep the rain off I wore an ex-army plastic gas cape. I came off the bike twice, both in the vicinity of Cambridge. One time, there had been a little rain, which made the road greasy. So I got into a slide, and slid sideways about forty feet. Luckily, neither I nor the bike were hurt. The second occasion was also in the rain. Again I came off, but this time I rolled myself into a ball, and rolled along what seemed a good fraction of a mile until I stopped. Luckily I didn't hit anything. I bought a similar bike in Australia, and my first wife, Margaret, rode pillion while pregnant, looking like a ship in full sail.

13.0.5 A VISIT TO DIEPPE

The weather being pleasant, Gertrude and I decided to take a day trip to Dieppe over in France. One could take the standard ferry from nearby Newhaven, or a hydrofoil boat from Brighton. The ferry was much cheaper, but took several hours to get to Dieppe. Arriving at the Railway station in Brighton, we found that the train had left, so we decided to motorbike over the Seven Sisters, cliffs between Brighton and Newhaven. I thought we had arrived early, and hied myself off to the men's room. Gertrude found the boat was about to leave, and sent someone in to unseat me. On board we found a sizable fraction of the passengers used this as an opportunity to get drunk, there being no tax on liquor. In Dieppe, we visited the cathedral and castle, but I found the most intriguing objects were the pissoires, which had not yet been done away with. With some difficulty Gertrude convinced me that one of them would not make a suitable garden ornament. We had coquille St. Jaques (scallops) for lunch, purchased six bottles of wine and a bottle of liquor, and returned to England. The Suzi carried the two of us, the wine and the liquor over the seven hills back to Brighton with no trouble.

13.06 BARN DANCES

The barn dances were a unique country dance get-together. They always had a live band. One such was put together by members of the university physics department, and went by the name of "Maxwell's Demon", accompanied by a delightful picture of the appropriate imp painted on the drum. The venue was generally a pub. I well recall one such, where we danced in a hall upstairs over the bar. The floor must have been weak, because it leaped up and down with the dancers. I was afraid we might well end up inadvertently in the bar as the floor collapsed. It did not however, but the motion of the floor made it a delight to dance on. When not dancing, one sat at a table holding on to one's beer. If one did not hang on, the glass would walk across the table with the vibration, and fall on the floor. The dances had fixed measures, and the caller blew a football whistle to finish. The band then wound up the dance at the next convenient point.

13.07 LESLIE HALLIWELL

One of my most interesting school-friends was Leslie Halliwell. He was an erstwhile member of the pig club, and had a similar distaste for organized games. However, his interests lay more in the modern than scientific subjects. He was bugs about movies, and attended one or two a week, even as a child. I would frequently accompany him and we would sit in the front row (cheapest seats). Also, come Christmas, we would attend the pantomimes, five or six in one season I recall. There is something about English pantomimes. It never bothered me as a child, but since I have lived in America I have some difficulty explaining why the "principal boy" is a girl, and why the "dame" (the comedian – a prized part) is a man. This is why it never bothered me to appear as an old lady on television when I was introducing the "Monte Python Show". Anyway, we both went to Cambridge, but after graduating, Leslie ran a movie theater for several years, which he recalls in his memoirs "Seats in All Parts". He then went on to work for Granada Television, and ultimately was their buyer for movies from Hollywood. His experiences there would fill a book.

He wrote a guide to the movies, which for many years was the standard if you wanted to look up the plot or characters in a movie. I sought him out one time when I was in London, and he showed me around his office, including a model about six inches high of the gorilla "King Kong". He told me this was the one used for the animation sequences in the movie of the same name, which he had inherited from somewhere. He had married and lived on Richmond Hill. We went to a play that night in the old Savoy Theater, on the recommendation, culled from the London Times, of a review written by another school-friend, Irving Wardle. The theater, which was where Gilbert and Sullivan's Savoy Operas were staged, burned down later but has since been rebuilt. It was much smaller and more intimate than I had thought. I never saw Leslie again – he died the next year.

13.08 SOUTH AFRICA

The weather in England. from January through Easter, was consistently miserable. If there was not a steady drizzle, it was overcast. I do not believe I saw the sun but once in this period. I was getting more and more depressed. When I lived in Lancashire under such conditions, it didn't bother me at all, because I assumed the rest of the world was just the same. In Brighton I pined for the sunny skies of South Carolina which I was now used to (though not the unbearably hot summers). The interesting thing is, the rainfall in Carolina is about the same as in Lancashire – but it falls all at once in immense thunderstorms instead of the continuous drizzle, and "Scottish mist" which pervades the air of the Northern Counties of England. Anyway, I had met Friedel Sellschop of South Africa the previous year, when he gave a talk at the University of South Carolina. He had offered me a visiting professor position at the University of the Witwatersrand during the forthcoming sabbatical, so I wrote to him and arranged to visit for the summer instead of staying in Brighton. This was the South African winter, corresponding to the English summer. The day I left Sussex for South Africa was bright and sunny – a perfect English spring day. I could see cricketers defending their wickets, it seemed all's well with the world – so I cursed myself for leaving. However, when I returned four months later to Brighton, and said to my colleagues, "I left just as the good weather was coming in" . "Oh no" they said, "It went back to rain the following day."

13.09 JAN SMUTS BOULEVARD

The first time I was in Johannesburg, I lived with a family on Jan Smuts Boulevard. The head of the family was a retired printer, but he was very sick. His wife ran the household, and had a live-in maid, and a girl who came in, as well as a wash-lady who took and washed the clothes. This business of having servants was new to me. I was used to doing everything for myself. I felt very guilty when I came home and found my corduroy pants washed and ironed. Guilty, but pleased – it was very nice. However, I talked about it to the lady of the house who, of course, had grown up with this situation. "Well," she said, "the wash lady has a family with kids. There is no social security or anything in South Africa. If I didn't employ her, she would starve. Really starve. As it is, she performs a useful service doing the wash, she feels she is a valuable part of society, as she is, and her kids don't starve". After this I felt very ambivalent. Clearly, there is something wrong with such a system – but if you must live with it, better enjoy it. What do you think?

I recalled seeing Jan Smuts with General Eisenhower when I was in Cambridge just after the war. They were watching the boat races on one side of the river Cam, and I was on the other. It is a narrow river, so they were actually quite close. So narrow in fact that the race consists of the boats chasing one another, since it is impossible for them to pass. The objective is for the prow of the rear boat to bump the stern of the front boat. Both boats then pull to the side, and exchange places for the race the following day.

13.10 LEARNING A CLICK LANGUAGE
For various obscure reasons, I resided in a girls dormitory in WITS (the

University of the Witwatersrand). This was very nice, I could observe these attractive girls from a distance (none of them would have anything to do with me!). One of the ladies serving in the cafeteria line took pity on me and said, "Why don't you come down earlier for lunch and sit over there," pointing. I did this, and found I was sitting with a group of women, mostly black, but a few white. This was, of course, illegal because of apartheid, but evidently a blind eye was turned on this situation. I talked to the black girls, and found they were much less interested in fighting apartheid than they were in getting a good education and going back to their home kraal, where they would teach. The white girls were there to learn a Khoisan click language – such as Bantu Anyway, this caused great hilarity, because a black girl would say something in Bantu, with the clicks being a beautiful part of the enunciation. Then a white girl would try. She would say something, then click, then go on. The black girls would be in stitches.

13.11 DOWN A GOLD MINE

One day Friedel asked me whether I would like to go down a gold mine. I jumped at the opportunity, then thought twice about it. The mine, in the East Rand, was well over 11,000 feet deep. Two miles down. Claustrophobia. All that rock above you. Anyway, away we went. The cause for going down was to recover the equipment used in the neutrino experiment in which Friedel and Fred Reines had first measured the natural neutrinos from the sun. The mine head had this gigantic wheel – far larger than those I recall at the coal mines of Lancashire. The cable went down something about 5000 feet, but its weight prevented anything deeper. The cab in which you went down had two levels. We were in the lower. It drops very rapidly, and has open sides, so you can see the shaft going rapidly by. The speed with which it drops with humans aboard is nothing compared to that when the ore is being raised, and it has happened they forget, and drop the people too fast, in which case they get flattened as the elevator stops. At the bottom, you get out, walk some distance and get on a second lift. The machinery for this lies in a cavern hollowed out in the rock. Down again, and yet a third lift. By now it is very hot, and in fact the principal cost of gold-mining is the air conditioning necessary to work in the galleries. The neutrino equipment lay behind a huge steel door. The mine periodically flooded. When it did, the experimenters came out, and locked the door behind them. When the flood subsided, they would return.

Our guides asked if we would like to visit the rock face, which was now some thousands of feet lower than the neutrino gallery We said yes, and descended on a rail car which went down at a very steep angle. The gallery at the rock face was being shored up when we were there.

I picked up a large chunk of the ore while 11,000 feet deep, which I still have. It looks very pretty, but it is fool's gold – iron pyrites – which you can see. Such ore gives less than an ounce of pure gold per ton. Mining was very unhealthy in the old days – most miners did not last above their thirties, dying of silicosis. They now have this beat. One lot of miners goes down, sets and explodes the charges. The next lot goes and wets it all down, thus eliminating the dust that killed people. The third lot loads the ore and transports it to the surface.

Emerging from the mine, we put on very heavy top coats. It was chilly out and miners would get pneumonia unless they kept warm.

The mine-work is performed by Africans from all over the southern region of Africa, each of which has their own language, so there is a special mine language, which is taught them after they arrive. It has very few, but very necessary words, such as, "Watch out the roof is falling," etc.

13.12 A DIAMOND MINE

Having been down a gold mine, I wanted to see a diamond mine. They are very different. Strangely enough, the diamond mines are not too far from the gold mines, being near Pretoria, although there is no geologic reason why this should be so. A diamond mine is simply a hole in the ground. Aeons ago, a volcano blew its top, and erupted, shooting out molten rock. Ultimately, the shaft or hole through which the lava erupted solidified to form the volcanic core composed of kimberlite, or blue ground as it is called. This is where the diamonds lie. They grow far underground under enormous pressure and heat. The eruption of the volcano shoots this stuff to the surface, cooling it, and after the top of the volcano wears away, the original core looks no different to other land nearby. The Cullinan mine, which I visited, is where the largest diamond in the world was found just below the surface, as big as your fist. It was given to King Edward the Seventh as a birthday present, which South Africans later regretted. Anyway, if you visit the Tower of London, you will see bits of the stone in the scepter, and the Imperial Crown.

Unlike the gold mine, you may not carry samples of the ore away with you. This is because there may be a big diamond inside. Very unlikely, but you never know. In fact, it is illegal to carry uncut stones in South Africa, because they were undoubtedly stolen. Friedel had a "License to carry uncut stones", which was a big deal. The blue ground, after crushing, is placed in a dense organic liquid. The ground floats off, and the diamonds, and a lot of other stuff, remain. The diamonds are placed in water and are washed over wax tables. The diamonds stick to the wax, and everything else goes on. Why they should stick to the wax, I have no idea. Anyway they are later scraped off the wax. An uncut diamond looks very uninteresting – like a piece of sandblasted glass. But after cutting and polishing, it is a very different matter.

The biggest diamonds are found close to the surface, because the diamonds dissolve going up when the pressure is suddenly removed, and the volcano blasts off. The diamonds getting to the surface first have had less time to dissolve, and are therefore the largest.

13.13 ISLA DU SAL

On my first trip to South Africa I left from England. I went to the South African consulate to get my visa, but they said, "Oh, this passport is no good – it lasts only a month after you leave South Africa, and regulations say it must be at least three months". Past experience had taught me it would take a long while to renew, and I thought maybe I would get a British passport, being in London. Having dual citizenship, I could do this. However, I thought I would give the US embassy a try, and went there, a magnificent building in Grosvenor square, with the golden eagle on top. I spoke to the consular assistant and explained my predicament. Much to my surprise, he said, "Oh, no problem, come back with a passport photograph and we will give you a new one in a couple of hours". I was astounded. He also told me the nearest place to get a passport picture. Sure enough, it only took a couple of hours.

One could not travel directly to or from South Africa at this time because of apartheid, nor could one fly over the African mainland. The airline had had an arrangement to stop over at a rather weird small island off the West Coast of Africa belonging to Portugal called Isla du Sal. There was absolutely nothing on this island except a tin hut, the landing strip, and facilities to gas up. It was very hot. I wonder what has become of it now?

We stopped off in Nairobi one time. It was a lovely day, and flying in we could see the snow covered peak of Kilimanjaro, poking up through a flat cloud layer.

13.14 RIO DE JANEIRO

The second time I visited Johannesburg, I left from the USA. You still could not fly directly to South Africa from New York, because of apartheid. So I had the idea to go via Rio de Janeiro. Apart from the fact I had always wanted to visit Rio, it lies eastward from the USA toward South Africa so there is a time advantage – it splits the jet lag into two. Flying down there I was lucky to take one of the few planes which breaks the journey. The plane I was on from Miami stopped first at Caracas in Venezuela. Our next stop was Manaus, at the confluence of the Rio Negra and the Amazon. It was dark when we arrived, but we flew low over the river, so we could see all the lights. Manaus was a thriving town with an opera house in its early years. I once had a school teacher who was interested in my book send me a card from Manaus, showing a kid holding a ferocious looking fish with big sharp teeth. It was a piranha! We flew on and landed at Brasilia at dawn. A very interesting modern city plunked down in the middle of nowhere with some odd looking tall buildings.

Then on to Rio. A colleague had told me I could easily obtain accommodation in Rio, but everyone around me in the plane assured me this was not so. A guy directly behind said he was an Australian, working for the state oil company. His wife was away, and he suggested I stay with him. This sounded like a good idea, so this is what I did. I thought he would have a small house in the suburbs, but it turned out he had a palatial apartment overlooking the sugar loaf, the big mount in the harbor. I had a good time in Rio, visiting the big statue of Christ, and ascending the sugar loaf by cable car. I also went to a night club with the beautiful coffee colored girls, and almost got washed away swimming from Ipanima beach. My benefactor took me sailing over to the Rio Yacht Club, to which he belonged. There were orchids in the trees around the clubhouse. It was all quite fascinating. Three days later, I went on to Cape Town, flying low over Table Mountain, and then to my final destination, Johannesburg.

13.15 CAPE TOWN

Whilst in Johannesburg, I had an invitation to give two lectures at Cape Town. One was on my research topic of ion solid interactions, the other on string and sticky tape experiments for kids. I flew down to Cape Town though I would have much preferred to take the famous "Blue Train", one of the few really elegant trains left in the world. It being midwinter, the weather was perfect, and I took a tour of the city, and the cable car up Table Mountain. Watching people take off to hang glide down onto the flat plain below proved fascinating. The up-draught of air over the edge of the table kept them flying to and fro for a long time. Walking through the Indonesian district by night I heard a familiar sound – even before I saw it, I thought it was the snap of Mah Jong tiles being laid down, and on going closer, I saw this to be true. The Indonesians had been brought back when Cape Town was a Dutch colony, used by sailing ships en route between Holland and Indonesia – the spice islands.

The Cape of Good Hope is the site of the famous "Flying Dutchman" ostensibly a place of storms and bad weather. I went out to the end of the cape (not Africa's southernmost point) but it was a beautiful day, and one could look down on the breakers from the high cliff above.

13.16 SUN CITY

My boss, Friedel Sellschopp, had promised to take his young son to Sun City, because all his friends had been, but he had not. So we went to a small game park, and also Sun City. This remarkable edifice is a miniature version of Vegas or Reno. The Nederlandse Hervormde Kerk, the regional church forbids all gambling, and, of course, you went to jail if a black person had relations with a white person in the Republic. No such restriction existed in the native homelands however, such as Bophutoswana. Some enterprising entrepreneur realized these were just the conditions which existed between California and Nevada. He had convinced the homeland government of this, and Sun City came into existence, and was very popular. I wonder what has happened to it with no apartheid?

13.17 SAFARI – KRUGER PARK

I could not leave Africa without taking a Safari. So I did – at Kruger National Park. It was a bus safari, but well worth it. En route, we stopped in Pretoria, viewing the school where Winston Churchill had been imprisoned after having been caught by the Boers, when he was a newspaper correspondent. We also visited the tunnel at Wasserfall Over. This train tunnel is mentioned in Churchill's memoirs, since it is where the railway descends from the High Veldt to the Low Veldt, the two plateaux. All that I recall about this is that a monkey, sitting high up on the wall of the cutting, threw a big rock at me, luckily missing.

We stayed in "rondavells" – round huts similar to those the natives use. Since it was winter, the grass was low, and we were lucky enough to see a pride of lions. Moving on, we passed a bunch of ladies picking reeds for roofing. Our bus driver said something to them in an African dialect, and they laughed. On asking him, he told us he mentioned we had just passed the lions. "It didn't bother them," he said, "So long as they stick together, the lions won't attack". I also found out what elephants eat in the wild. I saw an elephant wrap its trunk around a small tree, and pull it up. It then put the leafy top of the tree in its mouth and chewed – almost like a lollipop.

13.18 THE VAAL RIVER

Friedel Sellschopp had a vacation home he shared with his family, on the Vaal river, so we went there for a weekend. It was a fascinating experience. A dam just below his house made the river quite wide there, suitable for sailing etc. The house had a large central area suitable for eating, cooking etc., off which were a series of rooms. Each room belonged to a member of the family his brother etc., so they could leave their things there.

I saw cochineal insects on a plant. If you put your finger on a leaf you crush the insects, producing a bright red stain. This used to be the only red food dye available when I was a child. I was outside with Friedel when we heard the cry of the sea eagle. "You will return to South Africa," said Friedel. "Impossible," I said. "If you hear that cry, you will return," he continued. Well, he was right, I did return, many years later.

13.19 RETURN TO SOUTH AFRICA

I returned long after apartheid had ceased. I came back for a meeting of physicists interested in "Ion-Solids Interactions". Ken Purser and his new South African wife also attended. The same hotels in the hill district of Johannesburg which had been so segregated previously were now completely integrated, with black families staying there. All in all, superficially it seemed to me integration had occurred with very little friction.

The first night we went to a festival in a nearby kraal. They put on a drumming demonstration which was quite magnificent. They tune the drums by holding them to a fire, and the drumming is unbelievable, beating four against five, etc. WITS really did us well, and the meeting was well attended. Friedel died last year, so this may be the end of such gatherings.

14 RUSSIA, MATERIALS RESEARCH WITH THE ARMY, BOSTON, AND MY MOTHER'S DEATH

14.01 ARMY MATERIALS RESEARCH

The air Force, and, I believe, the Navy, offered research positions for faculty in the summer. This was a good idea, bringing new minds to bear on old problems, and teaching faculty what went on in these mysterious military institutions. In 1986, the army decided to follow suite, and I was one of the first participants, being assigned to Watertown Arsenal near Boston. This proved no big advantage, because the army was undecided whether to close down the research there altogether, which I gather they ultimately did, and the site is now a shopping mall. This is the only such institution I have been to where everyone was older than I. Nevertheless, they were working on some interesting problems, especially computer problems of plasticity in metals. I had taken a portable terminal an IBM "silent 700" I think, with me. At that time this was a big deal. Essentially, the device consisted of an electric typewriter connected to a very slow modem, into which you plugged the telephone handset. It is difficult to believe that most telephones at that time were permanently attached – no plugs. Anyway, since the army had a dedicated phone line, it was easy for me to interact with the computer back in Columbia. The results were typed out on the typewriter – computer, which had a thermal printout. I also worked on some of the army "finite element" computer problems, which were very new and interesting then.

14.02 CENTRAL SQUARE

The guy I was working for had graduated from Harvard, which gave him the benefit of using their list of accommodation. We pursued this, and I picked the cheapest digs. It turned out I was renting from an individual in the Greek mafia, an attic in a house in the area near Central Square. The principal problem arose because it was a very warm summer, and the attic was boiling hot. I bought a window fan, and after a while, Gertrude came and stayed with me for a few weeks, putting up with the heat. Still, being active in folk dancing and Bell-Ringing, I found much to do. I still have my MIT folk dancing T shirt. It reads:

$$((P^2W)^2W)^2(P^2WP^3W^2)^2$$

This is a dance, and means "polka twice, waltz once, repeat, waltz once, repeat the whole – etc., etc. They also had a Hungarian dance on a boat which toured the Boston harbor – it was delightful, with live Hungarian musicians.

14.02.1 CHANGE RINGING IN BOSTON

One of the benefits of living in the Boston area was the opportunity for excellent change ringing. Furthermore, one could go folk dancing every night, whereas once a month was about it in Columbia. The principal center for ringing was the church of the Advent, close to the esplanade. You could find any kind of ringing you wanted there, from rounds to spliced surprise. At the same time, they were finishing reinstalling the bells at the old North Church (Paul Revere rang there). Every July fourth, the Boston pops played Tchaikovsky's 1812 overture on the esplanade, with real cannon. Someone pointed out to conductor Fiedler they could have real bells, too, from the Advent nearby. So it was that I rang with the Boston Pops. It worked like this. A man with a walkie-talkie on the esplanade said, "Now". This was repeated in the tower by a guy up there receiving with a similar walkie-talkie, and we rang. At the correct time, "Stop" was relayed to us, and we quit.

A reenactment of Paul Revere occurred at Old North Church on the anniversary of his ride. A man dressed in a Paul Revere outfit was supposed to light a lamp in the cupola on top of the tower. Actually, he climbed one floor up where there was a switch – they had installed a candle-like fluorescent bulb at the top of the tower!

14.03 MOTHER'S DEATH

I was called to the phone one morning, and Christopher, my son, who was visiting my old home town Bolton, told me my mother had died the previous night. This was not too surprising, she had been in ill health for quite a while, however the circumstances were definitely unusual. Earlier, my son had told my mother that he had become engaged, and that he intended to bring his new bride over to England on their honeymoon. My mother was very anxious indeed to meet his wife, and sure enough, they came over, and spent a day with my mother in the nursing home. She passed away peacefully that same night. There can be no doubt she stayed alive to meet my son's bride, Karen. I took the next plane over, but it was a great help that Christopher had made all the arrangements for the funeral. There were few at the cremation ceremony, but it was nevertheless moving. The church of England minister gave a short address, and the coffin was drawn back behind the curtains. The ashes of both mother and father were cast in the same spot.

14.04 THE LAKE DISTRICT

The day after the funeral, I took Karen and Chris on a trip to the Lake District. This was a family favorite at the beginning of World War II, because it was easy to get to, by bus or train. We went to the Lake Windermere train terminal, walked down to the lake at Bowness and took a trip in the rain by steam boat up the lake and back. It had changed little since 1939, and was beautiful, peaceful and tranquil, but inevitably it rained – just showers, as it always does – it is the reason the women have such beautiful complexions there. The Lake District was a favorite of my first wife Margaret, she had worked as a waitress there in her summer vacations.

14.05 RUSSIA

My first visit to Russia followed on a physics meeting in Denmark. I flew to Helsinki, where I was shown around by a friend from the USA, Carita Barr, who happened to be at her old home in Finland at the time. We saw the monument to the composer Jean Sibelius, and had tea at an outdoor café. Then I took the USSR train to Moscow, bypassing St Petersburg (still Leningrad then I believe). The train was from the early thirties, and the track was miserable. It shook so much, I had difficulty urinating, but the toilet was locked when in the station – a trying situation. I shared a compartment with a German salesman. He showed me how to fix the bed. The top sheet was essentially a sack with a hole in the middle, into which you stuffed the duene, eiderdown, whatever you call it. At the border, a soldier came in with a small ladder upon which he stood, and pushed up a trapdoor in the ceiling, to look around and check no one was being smuggled in. I am not sure why one would want to get smuggled into Russia, but there you are.

Arriving in Moscow, I took a taxi to my hotel, which turned out to be the one Lenin had occupied on his return to Moscow in 1918, or whenever it was. I do not believe anything had changed since. It overlooked Red Square, so was very central. Much to my surprise, I got bitten by mosquitoes. I was there as a tourist, but thought I might see one of my Russian colleagues, so I went downstairs and asked the lady at Intourist, thinking the chances of contacting this guy at Dubna were slight. After about an hour, I got a call that this individual would be downstairs the following morning, and we had a delightful discussion. I did all the touristy things – saw the Kremlin, with the Czar Kolokol – the largest bell. (cracked) and the largest cannon (never fired – it would explode) but the old churches, slowly being restored were delightful. Three days later I returned to Denmark.

Prof. Tulinov had indicated he would invite me to a conference to be held in Moscow – but the invitation never came. There was clearly something going on behind the scenes, but I never found out what it was. Anyway, about two years later I received an invitation to a meeting in Moscow, and went. I flew into Moscow airport, and they lost my luggage trolley. The guy who was supposed to help me find it got me in a corner, and asked if I would like to exchange some dollars at a very good rate. Having heard of the hazards of doing this I said no, but this was probably a mistake. After all what could they do to you? Anyway, I stayed at the hotel connected with Moscow State University. It was very comfortable, but shook as if by an earth quake at one point. I recall I took a walk down by the river nearby, where a lady was picking flowers. She made some remarks which of course I did not understand, so I intimated I was American, when she presented me with a bunch of flowers. This really impressed me. I gave a talk in the big hall at the university. Since I speak no Russian, they appointed a translator. I thought this would be very tedious, but it proved not so. I would say a sentence, and he would translate. If he did not understand, I would explain. Normally, I would simply have gone on, and everyone would be lost since they too would not understand. Furthermore, since I did not understand his Russian, I would ponder what to say next while he was talking – making my remarks much more cogent. Hence, it proved a good talk. The walls of the entry to the hall were covered with photos of the students. Unlike America, where each student has an inane grin, the Russian students were evidently told to snarl. Have you ever been snarled at by several hundred students? A curious sensation.

I paid a visit to Dubna, the equivalent of Harwell, the British Nuclear Research Establishment, some miles from Moscow on the Volga river. My hotel overlooked the boats plying up and down – it must be a delightful cruise – you could hear the music being played on board.

I was provided with caviar for breakfast every morning – not the best caviar, but still caviar.

15 CHINA AND JAPAN

15.01 FLIGHT TO TOKYO

I had always wanted to visit China and Japan, so when the opportunity arose with conferences at both, I grabbed it. I was invited to give a talk at the University of Okayama in 1990 at a conference on atomic collisions. It is a long flight from Columbia via Seattle to Tokyo – the Pacific is a two movie trip (i.e. they show two successive movies on the plane). And, it was a long trip from the airport to the city – where I went to the wrong hotel – but the people were very hospitable and helpful, directing me to the right place, even though I don't know a word of Japanese. I had a very small hotel room, with a one piece bathroom – the whole thing was cast out of one piece of plastic. A nice feature however, which I have never seen in America was that slippers and a bathrobe come with the hotel room. I arrived with severe nasal congestion, and somehow managed to purchase a bottle of antihistamine pills – I still have the bottle with full instructions in Japanese. I went to a Japanese McDonalds for dinner. There were pictures of hamburgers and such on the counter, so all I had to do was point.

15.02 SHINKANSEN TO KYOTO

The following day I took the subway from my hotel to the railway station for the Shinkansen – the bullet train – I got a subway map at the hotel and people helped me find the train. There were several women on the subway in the most beautiful kimonos, and made-up like you see in pictures, with the hairdos etc. They must have been going to some ceremony. The trip to Kyoto cost about $80. The train was not crowded, so I got a good view of Mt. Fujiyama – it had then no snow on and looked somewhat dull although the exact conical shape was just like the pictures I had seen. The surrounding mountain country was very beautiful. There were people everywhere – every little bit of land was used up. The train was going at more than 200 km/hour (139 mph) but the speed was not really noticeable. It was on time everywhere. We arrived at Kyoto about noon and I had a hard time finding the baggage room to leave my cases while I saw the town since the signs were not very meaningful to me. I wandered about the station, and found the stores had much the same goods as those available in my home Columbia! I had a delightful Japanese lunch – the dishes are displayed outside as plastic models, so I took the waitress outside to the window and pointed. The previous night I had had Chinese food by mistake since I couldn't read the restaurant sign!

I took the #5 bus to "philosophers walk" and visited the temple up against the mountains. Although Kyoto itself is comparatively flat, there are quite steep hills behind it. Wandering around the temple grounds, I was astounded to meet Oakley Crawford – 15000 miles to meet a guy I worked with at Oak Ridge National Laboratory. He was with a colleague from Tel Aviv. We had tea together at a most delightful place – with huge colorful carp or koi in a small pool. The owner fed the koi for us. Leaving Ashley, I visited one or two more temples – one with the largest bell in Japan (so they said). On my own, I got completely lost, and was helped by a local resident, who had no English, to find the bus – I used to worry about travel on my own in countries where I speak not a word of the language, such as China or Japan but it really amazes me how one can get by. Oh yes, I know "Domo aragato" that means "Thank you" in Japanese – in Chinese it is "Shi-shi".

15.03 ARRIVAL IN OKAYAMA

Arriving in Okayama, I took a taxi to the residence – I had the instructions how to get there, but the taxi driver (they all wear white gloves) didn't understand – so we stopped at a little house on campus, probably belonging to the guard, who was asleep. He woke and explained to the driver where I should go, writing on a small blackboard – luckily, as we got there someone I know was emerging – otherwise I might never have found it. It was very comfortable – but no carpets, only hard mats on the floor. Looking out the window next morning at 6:30 am I saw an old lady gardening on the tiniest plot outside. Participants from all over the world were housed in our dorm, and each suite had a one piece plastic bathroom, with a quick heater turned on from outside, and the typical deep bath. The floor appeared to be of green oilcloth, and of course, you took off your shoes outside. Passing some school kids on the way to the conference, I noticed the girls all wore identical navy blue gym slips, and the boys a dark blue military uniform. The meeting hotel (Okayama Plaza) was very comfortable, and very western.

I was told by Prof. Wong, organizer of the Chinese trip, to cancel accommodation in China – it had all been set up – but communication with the USA from Japan is not simple, because of the 11 hour time differential.

Finally I got in touch through my wife Gertrude. It was all very complicated, but ultimately, I got it sorted out. The conference was, of course, all in English, the world's lingua franca, so that was OK.

15.04 THE CROW CASTLE

I visited the "Crow" castle in the city. It had been completely destroyed by the US during WWII, but rebuilt in solid concrete, so it could not be knocked down again! However, adjacent was a small tower which had not been destroyed. It was a moon-watching tower – you could have a party and go there to watch the moon – a delightful concept. Wandering around the town with a friend, Carl Rau, I bought an interesting toy. It consisted of two tops which were spun on a small shallow concave wooden dish, The tops resembled Sumo wrestlers, the immensely fat men who endeavor to overcome one another. The tops bounce off one another until one is thrown off the dish. We were presented with a curious square wooden Saki cup. Apparently, this is the traditional way to drink Saki, the wood keeps the wine warm, just like a Styrofoam coffee cup.

15.05 THE TYPHOON AND TOKYO

I returned to Tokyo via the Shinkansen, buying a box lunch on board. The weather had been bad, with the result that I saw Fujiyama covered with snow. In fact, just as I arrived in Tokyo, a typhoon struck. Much as with a hurricane in the states, it was the rain which was so noticeable. The wind, though very strong, was not so bothersome.

I took a tour of Tokyo, visiting a fair and a Shinto temple. I learned how to distinguish Buddhists from Shinto. When preparing to pray, the Shinto clap, in order to wake up God. You get a very good view from the top of the replica of the Eiffel tower. It overlooks an interesting cemetery just below. The cultured pearl shop is a Mecca for tourists, who are also taught the rudiments of flower arranging. It bears a certain similarity to the beauty of symmetry in physics – such as the eightfold way, although it is the symmetry breaking which makes the flower arrangement attractive.

15.06 HONG KONG

I flew from Tokyo to Hong Kong. It may seem a rather roundabout way to travel to Beijing from Tokyo via Hong Kong, however, my travel agent assured me this was best, and since I had always wanted to visit Hong Kong, who was I to disillusion her? Suzy Wong from "the World of Suzy Wong" had always appealed to me, so I stayed in the place mentioned in the book. However, it is now an upscale hotel area, not the rundown bar region of the book. Hong Kong was hot and humid – just like home in South Carolina – luckily, they have air conditioning. I went down in the morning for breakfast, and was astounded to find no one in the restaurant spoke English. As a British possession, I had always thought they would speak a little English. Anyway, with a little hand-waving I managed to order a suitable breakfast, and it was delicious.

I had once been faced with a similar situation in Miami – a restaurant where only Spanish was spoken. Luckily I remembered that "Huevos" is eggs in Spanish, and that got me by.

15.07 TIGER BALM AND VICTORIA PEAK

I took a tour of Hong Kong, going up Victoria Peak, where you can look down on the skyscrapers and see where the rich folks live, and down to the beach named for the fact you cannot get down to the water without treading on someone. The Tiger Balm garden took my fancy, built by the entrepreneur selling the stuff. Every child seemed to want to have its picture taken sitting on an immense brass turtle. Is this considered lucky? The double decker trolley cars reminded me of my youth in England. I don't know anywhere else that still has them. Never have I seen such a density of people in a city – all very friendly however.

The city on the water, where everyone lives in a sampan, was quite fascinating. It had a weird name for China – Glasgow I believe! The gigantic restaurants floating on the water were quite unbelievable – I have since noted they have been featured in many movies.

15.08 BEIJING

From Hong Kong I flew to Beijing, and was met at the airport by an attractive young lady. She shepherded me through customs, and took me to my hotel. One curious feature of the hotel was the clock, on the headboard of the bed. It was telling the wrong time, so I reset it. However, a few hours later, it was still telling the wrong time. I finally worked out what had happened. No doubt the hotel company had got a good deal on beds from some foreign manufacturer. The foreign country had a different mains frequency, 50 instead of 60 Hertz for example. Since the clock came with the bed this would not have been noticed. So, short of exchanging the clock, they we stuck with it!

15.09 TRAIN TO JINAN

I took the train to Jinan the following day. Booking the train in America, I noticed an option was "soft bed". The hard bed proved to be a plank. The soft bed I had reserved proved very comfortable, and it was nice to watch the Chinese countryside drift by, so different from home, with the sampans and other boats as we crossed the Yellow River. The hotel in Jinan, which backed onto the "mountain of a thousand Buddhas", was very modern – they even took credit cards! I went down to breakfast the following day to find there were two possibilities. One was a typical English breakfast – bacon and eggs etc. – and the other was a typical Chinese breakfast. I decided to try the Chinese. I started with what appeared to be uncooked dough, with soy milk, which was a dirty white. The rest of the breakfast was similar, so I decided to stick with the English breakfast in future. I always believe in doing what the natives do, just to see what it is like to live like that – but one should only go so far. For example, one night we were provided with cooked sea slugs for dinner. These are a great delicacy in China, but I found them quite miserable.

15.10 THE MOUNTAIN OF 1000 BUDDHAS

At this time, Jinan had only just been opened to foreigners. Previously it was forbidden territory because of security considerations – it is not too far from the sea. The result was, we were the first non-Chinese many people had seen. So people would stop in the street and stare at us, as they might a performing bear or something. They had no animosity. One time I climbed the mountain of 1000 Buddhas with an old student of ours from Australia, Don Gemmel. The stone gate to the mountain was being rebuilt by stonemasons, and we stopped to talk to them, helped by a passing Chinese who wanted to practice his English since he was going to medical school abroad. Don is tall, and has red hair, so all this proved most interesting to the Chinese. They were all small – not one of them came up to my shoulder, and I am not tall. They had been imported to work on this gate, and lived in tents nearby.

A friend from the College of Charleston, Fred Watts took his family on sabbatical to Qingdao, (the seaport where they brewed the Tsingtao beer we all drank). His son, who was about nine, noticed that his head was patted by each Chinese who passed him on the street. This peculiar habit disturbed him. It turned out that he was red headed. No Chinese has red hair, and red is the Chinese lucky color – so the passerby patted him on the head for good luck. After learning this, he was less bothered.

15.11 CONFUCIUS TOMB AND MOUNTAIN

We climbed Confucius' mountain, and visited his forest and town. There I had a street vendor paint a sign "Wu Lee" for me. He did it with beautiful grand sweeping gestures. I was paying him when another vendor thought he was charging too much, so a big argument arose as to how much I should pay. I was not involved in the argument. It was there I met a relative of

Confucius. It turns out if you are related to Confucius, you can be buried in his forest. However, since Confucius died in 479 BC they must keep pretty close track of all this. I wonder how many Americans could trace their ancestry back that far? I had developed a pretty bad cold by the time I reached the mountain top, and our host, noticing this, called his secretary, who produced a brown pill from a bottle she had in a very large bag she had with her. I took it, and it certainly alleviated my symptoms. Back at the hotel, my roommate moved out, because my cough sounded so bad. I had gone with him downtown and purchased some liquor. Since there was no way I could work out what was in the bottle I had to guess. I believe it was plum brandy. Another help was China tea. Each day a thermos flask of hot water was placed in my room, together with what looked like little twigs. This was tea. You placed the branches in the hot water, and when they sank, the tea was brewed. It really helped with my cold.

I caught the train back to Beijing. It was something like Agatha Christie's Orient Express of the nineteen thirties. The "soft bed" I had been promised in my reservation proved most restful for my cold.

15.12 BEIJING – THE WALL AND TOMBS

Returning to Beijing, I visited the doctor at the hotel to get more pills. At least they delayed my symptoms till I got back to the USA. I took a trip to see the great wall. You can walk quite a long way along the wall, and I saw a train go by on its way to Moscow via the trans-Siberian railway. What surprised me was they had camels near the wall – originally for transport, but now for tourists. We visited the underground tombs of the emperors on the way back. The stone mythical beasts outside looked very odd – a real bestiary. There was a cafe named the "Ding Dong Restaurant". What a delightful name. I did not see a giant panda, in spite of all attempts to get us to go to the zoo. One night we had a delightful candlelit meal of Peking (or is it now Beijing?) duck. It turned out the reason for the candles was not romance but a power failure. The duck was so absolutely delicious a lady participant and I decided to view the kitchen. Going back there we found ducks everywhere. All the sinks were filled with ducks, together with the tables, and they were all over the floor. We came to the conclusion it was perhaps not wise to examine the culinary techniques in China too closely. Japan on the other hand was scrupulously clean – but they eat a lot of raw fish. Well cooked in China was my motto from then on. We were given a musical interlude with Chinese instruments. Particularly the one string fiddle, and the peculiar Chinese mouth-organ interested me. We also had an exhibition of odd juggling – including a guy who had a boulder on his chest smashed with a hammer whilst lying on a bed of nails.

15.13 RETURN

I returned to the USA via Chicago. As soon as I hit the USA, the symptoms of my infection reappeared stronger than before. It seems the pills only worked in China!

16 AARHUS

16.01 WHY AARHUS?

My last sabbatical was spent at the university of Aarhus, on the Jutland peninsula of Denmark. My wife Gertrude is from Denmark, and long before I knew her, I had intended to spend my first sabbatical there rather than Munich – but Munich came up with a better offer. Gertrude has innumerable relatives that permeate the heart of Denmark, and the research was congenial also, pertaining to ion-solid interactions, so it seemed an obvious choice. Nevertheless, I never did manage to learn Danish, which some say is a disease rather than a spoken language. My best effort was "fadøl" which I learned at a bar in the airport at Copenhagen and means "draft beer".

16.02 ACCOMMODATION

We left America in the fall of 1991. The problem of accommodation was solved by my sister-in-law Hanne Stengård Larsen, who had put an advertisement in the Aarhus newspaper that we would be willing to exchange houses, cars etc. for a period of several months. My thought was, why would anyone in their right mind move from the delightful countryside of Denmark for the heat, humidity and scrub of South Carolina? But it turned out several did, and we finally picked the Hammerholt family. They came over here before we went to Denmark, but in any case, the exchange was started without too much fuss. They had a house on the outskirts of Aarhus adjacent to a "coloniehave" enclave. This is an arrangement where the Danes have a small allotment in the country to grow things, and often they have a very small cabin where they can spend the night. They grow the most amazing and delicious vegetables, and often a group of the growers will have a party – at Christmas, for example. We enjoyed living Danish style. I found the food really agreed with me, though I never did take to the raw herring – unless accompanied with several small glasses of high proof Aalborg akvavit. Gammel Dansk, (a famous liqueur meaning "Old Danish"), I found too bitter. I was glad when I had had enough, as they say. We inherited a beautiful car from the Hammerholts. We used it to travel all over Denmark, and I got quite used to driving on and off the ferries. Gertrude saw more of her own country than she had ever done before. What we did not know was that the car was part of a rental agreement, so the Hammerholt may have had a big bill when they got back. Under this agreement they always had a new car, so they never checked the oil – it was always OK. In driving our old car, they did the same with the result that it ran out of oil, and the motor packed in on the way to Charlotte one day.

The Danish countryside is flat, but the buildings are very interesting. The "crow's nest" or half-timbered buildings which they call "binningswerk" dot the countryside. We also found the flea markets (loppemarkt) fascinating. They were held inside in an atmosphere redolent of tobacco smoke so dense you could cut it with a knife. It was reminiscent of old English smog.

16.03 THE UNIVERSITY OF AARHUS

Aarhus, the second largest city in Denmark, has a university set on a hill overlooking the harbor. It is mostly of new, red brick construction, with the physics building seven stories high, located in a sylvan setting. I was given an office to share with one other individual, and the research lab was just down the corridor. I ended up assisting Eric Bøg in his work on electron emission from crystal surfaces. I also looked at electron scanning tunneling microscopy, but in fact, my research there was not very profitable.

16.04 DANISH HIGH SCHOOLS

On the other hand, I was interested in the high school (gymnasium) system there. They had just revamped their physics curriculum, and I found this fascinating. Unlike America, with its cumbersome bureaucratic method of changing course structure, Denmark was small enough for all the high school teachers to collaborate in setting up a new curriculum They had agreed on a core curriculum, and then how each teacher could decide in what way to branch out from this into other more specialized areas. All in all a very good idea.

16.05 PHYSICS OLYMPICS

I suggested to the high school that they might institute a "physics Olympics". I had done this many times very successfully in America. The idea is that, on a scheduled day, the students are divided up into teams of three or four, and then compete in specific experimental events. One is often the "egg drop" competition. The objective is to wrap, or prepare an egg in such a way that when dropped from a fixed height, it reaches the ground in the shortest time, without breaking. You are given string, sticky tape and paper. All kinds of bizarre arrangements are used.

The event was scheduled just before the Christmas season, because that is when the weather is at its most dreary and miserable. The different experiments really cheered things up. One aspect which surprised me was that the senior students were allowed to smoke and drink beer in the school. Anyway, the Olympics was a great success. I had emphasized that an integral part of the Olympics was that the media should be involved. The Danes are very laid back, and the thought of other people being involved never entered their heads – but I made a strong point of this, so the other teachers worked with the newspaper and radio, and I worked with television. As a result, I was interviewed on Danish TV – a very interesting experience. I was "texted". Since I spoke no Danish, they interviewed me in English, then when the item was broadcast, a translation into Danish was flashed across the bottom of the screen. The students were very pleased to be on TV – it gave them a sense of importance. It also enlivened what was a drab period in their lives – and I enjoyed it very much.

16.06 THE FOLKEDANS HUUS

My principal interest outside the university was folk dancing. One evening a week, a get together was held in a building called the "folkedanshuus". We formed squares, or longways dances to the accompaniment of a violin and accordion. We were instructed in the dance by a capable individual who knew no English. So I rapidly learned what "to heure" and "to vestre" meant. We even learned the quadrille, "Les Lanciers", still popular in Denmark. After two or three dances, chairs and tables were put out, and we sang folksongs together. Then coffee, clear tables and chairs away, and on with the folk dancing. They had a fantastic Christmas party, where we all marched around the Christmas tree singing carols. We did the same thing at the house of Hannah and Annes Hansen. Gertrude and Hannah are sisters.

16.07 CHRISTMAS IN DENMARK

The weather being so miserable, indoor Christmas events form an important part of life in Denmark. We were enthralled by a television program which came on for fifteen minutes each evening from Advent to Christmas. It was called the "Julekalendar", and involved the antics of three Julenisse (Christmas elves), and a farm couple on west Jutland, or Juland, together with a nefarious character from Copenhagen. The inhabitants of this area, the "Jusk" have a dialect difficult to follow, so they were "texted" as I had been as a foreigner! The three nisse, and the two farm people (plus the Copenhagener) were played by the same three performers – the "Nattergale" (nightingales). It was a delight to me, because the nisse spoke a mixture of Danish and English which was quite hilarious. The plot was vague and difficult to follow, but basically the nisse were trying to recover an old book of magic spells to save the life of their sick relative in Canada. Suffice it to say, everybody enjoyed this farce, and watched it every night. We had to go to a Christmas party at the university, and thought we would miss this one episode. Not so. Come time, all the people at the party gathered around a nearby television, and watched.

We went carol singing, and I took my accordion. Passing a nurses' dorm, we ran into a "nissepia" (girl elves) party. All the nurses had put on "nissehue" for the party. These are rather like the red pointy Santa Claus hats in America. Luckily it was a bright but cold night, perfect for caroling.

16.08 THE BOHR INSTITUTE

I had been invited to give a talk at the Bohr Institute on Blegdamsvej, Copenhagen. It proved a most interesting experience. Historically, the Bohr Institute was a center for theoretical physics from the time it was founded until the end of the century. When I arrived I passed Ben Mottelson (whom I had met at Los Alamos, and who won the Nobel Prize jointly with Aage Bohr) taking down a book from the library shelf. The people who invited me showed me around, and we ended up in Bohr's sitting room, which had a fireplace, and generally relaxed air. I was then passed on to another researcher, and we ended up in Bohr's sitting room again. I was handed on for a third researcher, and guess what? We ended up in Bohr's sitting room. Not that I minded – it was a very comfortable place.

17 RETIREMENT

17.1 Faculty Meetings

If you get the offer of six months' worth of salary free and clear if you retire, take it. That was the advice of my accountant, so I took it.

I found that I was busier after retirement than before, and I wondered how I ever found time for work. On the other hand, everything I did I wanted to do – and I avoided all the things I disliked. For example, I loathed faculty meetings. It was a great comfort not to have to attend them. These tedious proceedings dragged on interminably, interspersed by incidents where one faculty member went bezerk and attacked another. One such member emptied the trash can over a colleague in a particular argument, and I have seen members stand and make fists at one another, as if to start a fight. Each member had their own hidden agenda, which they were anxious to promote, and there was great tension, especially if the budget was cut. It was said that these fights were so vicious because the stakes were so small.

17.2 Distance Education Teaching

The semester before I retired, I put two courses on video tape. These courses were still being taught some ten years after I made them, and I feel I shall be teaching them long after I am dead. The purpose was to allow students in the boonies, way away from civilization, to get an education. They watched the tapes, answered the set questions, and took the tests, mailing the results in, except for the final which had to be taken at the university. These were then marked, and a grade assigned. Where my course differed from others was that there was a lab attached to it. I had great fun making up experiments which required a minimum of equipment. String & sticky tape came in handy!

One interesting occurrence followed this course. I was walking across the horseshoe one day when a middle aged gent greeted me heartily. He was, of course a former student who had taken my course in his teens. As a professor, this happens, so I shook his hand as if I had known him all my life. "I really enjoyed your course", he said, "I remember that bowling ball you had on a long wire. I don't remember why you did that, but it really impressed me". I still believe you have to interest the students before they learn anything!

17.3 Plays and things

I have a bad memory. This seemed to preclude an acting career. Unfortunately for me, I enjoy acting, and have a powerful voice, as all know. My first connection with the stage was at school, where I was recruited for a walk-on role. I was one of the people who carried off the bodies in Hamlet or Macbeth. As anyone seeing these plays knows, the stage becomes littered with dead bodies towards the end of the performance, and each body is loaded onto a bier (not beer) and carried off. This seemingly simple act was, in fact a real problem. The reason was that the stage at our school was too narrow. This meant that the bier-stretcher carrying the body had to be poked through a window (invisible to the audience) at the side of the stage to remove it. If the "body" was not nippy in getting off the bier, he stood a decent chance of falling twenty feet onto the ground, since the stage was on the second floor (first floor English style). Luckily, this never happened in my apprenticeship, but the possibility was still there. This was the end of my performing career, until, in Australia, we put on a series of play readings. This was perfect for me – I didn't have to memorize anything, and I could give vent to my talents. You should have seen me in Moliere's The Miser" and "Waiting for Godot". We gave two performances of the latter. In one, the audience laughed themselves silly, and the other no one cracked a smile. Well! In America, the Unitarian fellowship to which I belonged acquired a small group of play-readers. Everyone attending got a part. We ran like this for at least twenty years. Sue Folk and Carol McAlpine also helped run it.

I was sitting at home about seven thirty one evening, in 1996 or so, when the phone rang. It was a director of Town Theater, a lady who told me that one of their actors had vanished. He took the part of a Lancashire businessman in the Agatha Christie murder mystery "Appointment with Death", and they needed a replacement. What happened to him they did not know. So I was asked, would I take the part? Of course I said I would, but I had this memory problem. "Never mind," I was told. "When can you get here?" "In about half an hour". Well, I did my best to learn my part, with my wife's help. But I am hopeless. Anyway all went well until the time of the first dress rehearsal, when I just got stuck in the middle of my speech. Instead of getting mad at me, the director just laughed and laughed as I stood there, dumb. She said, "Just keep on going – make it up, and when you stop, the others will feed you lines and you will remember," – and I did just that. I do not believe there was one performance where I did not forget something, but it all worked out fine. I am very good at ad-libbing. Perhaps, had I known this earlier, I might have been an actor!

I thought that would be the end of my acting career, but I was invited to take part in a performance one more time, also at Town Theater and if there was one play I would like to be in, this was it – "Noises Off" by Michael Frayne. It was the bedroom farce to end all farces. The set consisted of a staircase leading up to a landing with three doors leading off, and downstairs, a window leading into a garden. I was a drunken burglar, the other members of the cast trying to remove my bottle so I could go on. The first act showed the performers trying to learn their parts while we learn about them. The whole stage set then rotated for the second act, so the audience now was behind the scenes, and it rotated back for the third act. It was hilarious. As usual, I forgot my lines – so much so that there was one point where I was stealing a jug and said, "How much is this worth? Five quid?" Except some days I would say five quid, and some days ten. The last night, the leading man came up to me and said, "I bet my wife you would say five quid tonight, will you do it?" so of course I said yes. Then his wife asked me, "I bet my husband you'd say ten quid, will you do it?" This put me in a bit of a quandary. In the end I said, "What is this worth, five quid? No ten!".

The play is based on everything going wrong, but the strange thing was, many things went wrong which were not in the play. The culmination was the leading man falling downstairs, which received applause every night. We had a professional come in to show him how to do it without hurting himself. It was a fantastic experience. Who would think I would get into the acting profession at 65 – well, anyway nearly get into it – scarcely the time to be starting an acting career. Again, my accent helped, but my memory has always been bad. I could not remember a name, and said to my son, "I'm getting old – my memory's shot." "No no, you were always like that!" said my son.

The culmination of my acting career was the herald in Lysistrata, the Greek comedy against war. Ours was a demonstration against the war in Iraq, and was held the day before Bush declared war. It was a reading, but we had an audience of four or five hundred, and I had to wear a two foot long plastic penis. Not bad for a 74 year old!. Ann Dreher directed. Many years before, she and I were the speakers against marriage in the only British Union debate that the South Carolina Columbia Unitarians ever had. We were both married at the time, but both divorced later. Oddly enough, the pro-marriage speakers were both single.

17.5 Physics and Philosophy

A year or two before I retired, we had a physics faculty member from Israel,

Lev Vaidman, who was an expert on the "many worlds" interpretation of quantum mechanics. At that time this was in limbo, but it has since suffered a revival. He started a joint seminar with the Philosphy Department, which proved most interesting. He had to return to Israel, so I took it over for a year or so. Then the philosophers inherited it. Lev had been in the International Physics Olympiad – but from the Russian side (his birth place)! He was Jewish, and this inhibited his acceptance at university over there. However, if you had been in the Physics Olympiad, irrespective you were sure of acceptance.

18 RETURN TO AUSTRALIA

18.01 SETTING OUT

One does not take a weekend trip to Australia. So, after I left to go to America in 1958, I really had little opportunity to return till I retired. Then, in 1996, I decided to return for a period of several weeks to revisit old haunts and see all the sights I had missed as an impecunious post doc. My decision was reinforced by John O'Connor, a friend of mine with whom I had worked while on sabbatical at the University of Sussex near Brighton, and who was a professor at the University of Newcastle in northern New South Wales. When I told him of my decision he said, "Why don't you give talks or workshops about your string and sticky tape experiments? I tell you what I'll do – I'm president of the Australian Institute of Physics and I'll put it in our newsletter that you will give talks for free for your transportation and accommodation in Australia". Much to my surprise, I received several offers by email to give such talks. This proved a great boon. Apart from paying for my stay, I met the most interesting people, and saw things I most certainly would otherwise not. Email proved indispensable. Since Australia is about twelve hours different from us, communication is difficult. But I would send an email, and receive the reply the following day. In the end, I gave about six talks, all over the place. Apart from Newcastle, I spoke at the university of Western Sydney, Sydney University, the University of Melbourne, Monash University, which is also in Melbourne, the Australian National University , and Flinders University in Adelaide.

18.02 HAWAII

Since neither Gertrude nor I had been to Hawaii, we decided to make a stopover there of a few days. Unfortunately, the flight from Atlanta to Los

Angeles on August 21 got delayed, so we missed the flight out. The airline
put us up in L.A., and we got a flight the following day. Because the delay
had been their fault, they put us in First Class. It is the only time I have ever
flown first class, and a flight to Hawaii proved a good choice – it is a very
long way. What I call a "two movie Flight". Tokyo, or Sydney, to L.A. is a
three movie flight. Arriving in Hawaii, we went to the hotel. They had also
mislaid our bags, so I used this as an excuse to buy an exceedingly bright shirt
and pair of swimming trunks. Being colorblind, I find jazzy colors most
attractive. I believe there has been discussion that Van Gogh was also color
blind, because he employed such vivid colors. We had a very relaxing time in
Hawaii – drove around Maui and inside Diamond Head, (which is the cone
of an extinct volcano) saw the pineapple plantations, broke a few Macadamia
nuts – boy, they are tough, and flew over to the big island of Hawaii, where
we watched the volcano producing a river of lava and walked over recent
flows, and through one of the tubes. These are produced by lava flows which
cool on the outside, solidifying. The inside runs out, leaving a hole through
which you can walk. Most interesting, I rang the new ring of bells at the old
cathedral in Honolulu, which was donated, somehow, by Queen Victoria, I
believe. The ringers were still making the ringing chamber habitable – it was
very hot indeed. The islands were taken over by unscrupulous American
entrepreneurs, who dealt with the royal family. US foreign policy is not always
attractive! I also took a ride in the Yellow Submarine – a real submarine which
descended about twenty or thirty feet. I presume they have water-filled ballast
tanks to make it rise or fall, but I was a little trepidacious, so I heaved a small
sigh of relief as we surfaced.

18.03 TONGA – QUEEN SALOTE

Anyone who watched the coronation of Queen Elizabeth will remember
Queen Salote of Tonga. She was a statuesque, ebullient, fun-loving woman,
who rode in an open carriage wearing her brilliantly colorful native outfit,
waving and laughing to the coronation in spite of the rain. I had always
wondered what Tonga was like, and purely by chance, our plane stopped off
there on the way to Australia. The island having the airport was rather flat,
covered with coconut palms, and had lots of what looked like thatched
plywood shacks on it. It turns out, unlike most Polynesians, the Tongans are
big people. The Tongan airline was there, and appeared to consist of a rather
small corporate jet. I bought a few souvenirs, and after taking off got talking
with some of the Tongans on their way to Auckland. The principle topic of
conversation was the king's weight (Queen Salote was long dead). He had
gained too much, and was on a diet, which he did not like. Anyway, it turned
out that he had offered a prize to the overweight Tongan who lost the most
weight. I forget the details, but it struck me I would like to live in a country

where the most important problem was the king's weight!

18.4 AUCKLAND AND SYDNEY

We flew in low over Auckland in good weather, from which we could see the beautiful lakes around the city. Then on to Sydney – quite a long flight. Flying into Sydney is always a delightful experience. The harbor, Opera House, bridge, cruise ships and yachts make for a fantastic view. Reminded me of my first air trip, in 1954, taking off from Sydney in a DC3, with a thunderstorm hovering nearby.

We stayed on Bondi Beach overnight. I had wanted Gertrude to see this magnificent beach, and in fact the beach is still gorgeous, but I am afraid the surroundings have gone downhill somewhat. Nevertheless it proved a memorable experience, walking along the beach in the late afternoon. We had been told to look out for muggers and such, but the hotel proved fine.

18.5 CAIRNS

The following day we went on to Cairns. As an impecunious graduate student, I had never been able to afford to see the sights of Australia – Ayers rock, and the Barrier Reef. I suppose Ayers rock is the Taj Mahal of Australia, in that once you have seen it, there is little else in that vicinity. On the other hand, the area around Townsville and Cairns has very much more there than the reef itself. We stayed at a small motel in Cairns, and the next day I went out on a gigantic catamaran, which traveled at high speed to a platform in the most prolific area of the reef in terms of colorful fish. We also went via the tropical rainforest sky cableway to Kuranda, through the mountains to the plateau behind Cairns, returning by a cute old railway over the valleys to Cairns.

We returned to Sydney, and stayed in Kings Cross, which resembles Washington Square in New York, being the most Bohemian area. It proved quite exciting, and not overly expensive. You could walk down to Woolamaloo from there (the name had deeply excited Ted Irving, a close paleogeologist friend of mine in Canberra). I gave my talk/workshop at Sydney University. We stayed one night close to the university, from where you could see the new bridge – not over the harbor, however.

18.6 UNIVERSITY OF WESTERN SYDNEY

When I lived in Canberra, there was no Western Sydney University, but now there is, and we stayed with a charming couple, he being a faculty member there, and, of course I gave a workshop – with a lot of interaction. He gave me the "Ethnic Balloon" idea. He also took us to a small, but delightful zoo, where we were allowed to hold a koala bear. The bear has sharp claws, so the keeper gave it a cushion, into which it sank its claws, and then we held it. Quite delightful.

18.7 CANBERRA

From Sydney, we flew to Canberra, where I rented a car – public transportation in Canberra being rather weak. We were met by Dr. Fletcher, who was connected with the Australian Physics Teachers Society. Canberra had certainly changed since I lived there. Previously there were no lakes, and now immense stretches of water lay between parliament house and the city center – what we called "Civic". I found the hospital where my son was born was now an old people's home. Nevertheless, much remained of the old Canberra. A most attractive flower show was going on while we were there. In addition to giving a talk at the University, I also gave one at the Questacon, similar to the Exploratorium in America. We drove up into the hills toward Mt. Stromlo and the Cotter valley. At the top of the mountain we came across a herd of kangaroos who regarded us with surprise before hopping off, and driving back via a radio-astronomy observatory, we saw a group of wild emus, flightless birds. Stromlo observatory has since burned down in a big bush fire.

18.8 ADELAIDE

From Canberra, we traveled to Adelaide. My cousin David met us there, though we stayed with the parents of the wife of our former Unitarian minister in Columbia SC. I had not seen David since leaving Australia, nor had I met his wife. We met with some trepidation, but it turned out that we got on very well, and he took us on a tour of Adelaide, and down to the beach. I gave a talk at Flinders University, which is a relatively new university, founded after I left Australia. The university was in the hills behind Adelaide. As usual, the teachers were eager and active, and I had a fun time. I found Adelaide a most attractive city. They pride themselves on never having been a penal colony.

One of the curious features of Australia is the railway system. Queensland had a narrow gauge track, New South Wales a standard track, and Victoria a wide gauge. I am not really sure why this was, but I think pride had something to do with it, since the trains from one state could not run in another. Anyway, when I lived in Australia, in traveling from Sydney to Melbourne, you had to change trains at the border, a considerable disadvantage in the middle of the night. Since then, they have installed standard track between the state capitals. Evidently this is true also in South Australia, where the track passed close to my digs, and I found three rails. One pair was the original track, and they had simply added a third rail to allow the trains from Victoria through.

18.9 NEWCASTLE

We had intended to fly from Adelaide to Sydney, then change planes, but had mistaken the day, so spent one more day in Adelaide. The trip from Sydney to Newcastle on the 18th of September was carried out in a small plane at low altitude, so flying up the coast we were able to see all the scenery, including the Entrance, where I had spent my first honeymoon, in cold weather. The coastline is quite rocky, for the most part, all the way up. making it interesting. As one might anticipate, Newcastle had been a coal mining town, but no longer. I gave my talk at the university there, and was given a tour of the surrounding area. The area behind Newcastle grows good wine grapes, the Hunter valley, so we had a convivial evening trying the wine.

18.10 MELBOURNE

We returned to Sydney, and then on to Melbourne. Ken Opats had me give a talk at Melbourne University. We went there by trolley car (tram) and we thought the track very shaky, but it turned out there was a minor earthquake as we were traveling. He was having a party the following night which proved very pleasant. We also went up the TV tower in the middle of Melbourne, the view from the top proving how extensive Melbourne has become recently.
I also gave a workshop at Monash University. This is at some considerable distance from the heart of Melbourne, so it took me longer than I had thought to get out there, but it has wonderful facilities.

18.11 SYDNEY UNIVERSITY AND RETURN

We returned to Sydney. Everyone should visit Taronga Park zoo, where I saw live duck-billed platypuses (or is it platypi) for the first time. We ferried back to Circular Quay, and went up the TV tower, from which you get a fantastic view over the harbor. There were few liners docked compared to when I first arrived in Sydney. That has vanished as a means of transportation, but there were cruise ships aplenty.
We left Sydney 4.45 pm on the 27 September, and arrived back in L.A. at 1:15 pm the next day. Of course, we crossed the date line, but I am still confused how long it took. Leave L.A. 3:10 pm and back in Columbia 11:55 pm – a very tiring trip.

Ron Edge

19 AAPT, IYPT, AND THE PHYSICS OLYMPIAD

19.01 THE PHYSICS OLYMPIAD

In the spring of 1985 I received a mysterious phone call (or was it a letter?) asking whether I would have time to attend the XVI Physics Olympiad competition in a city called Portoroj, on the Adruatic Sea near Piran in Yugoslavia, as it then was. I said I would be delighted, but I had never heard of the Physics Olympiad. Nor, it appeared had any of my friends. What it boiled down to was that the Olympiad was a physics competition for high school students in their senior year, and the US was trying to decide whether to compete or not. At that time, there were about eighteen countries involved, mostly in Europe, the competition having started in Russia, Moscow State University, although it was a Polish professor, Waldemar Gorzkowski, who was now in charge.

19.02 PORTOROJ

I flew to Trieste, rented a car and drove to Portoroj, which is a seaside resort, popular with Europeans. Booking into the hotel I was astounded to hear country and western music being played over the radio at the registration desk. Later, I had to go back to Trieste to pick up Arthur Eisenkraft, who was the other observer. It was common for the observers to review the competition, and return to their home country to suggest they participate the following year. It turned out there was one other country also observing – China. So we saw quite a bit of the two Chinese observers. One of them confided that he had been born in the US, his father then being on sabbatical, had returned to China, and never been out of the country since. Yugoslavia was enduring a violent fit of inflation, so we changed our money twice a day, receiving a rate ten or twenty percent greater each time. Meeting the leaders

202

from different countries was very interesting.

I decided to take a short trip to Piran, and entered what appeared to be a wide street. As I progressed, it got narrower and narrower, until I suddenly realized, the walls were narrower than the car, and were scraping the sides. I slowly backed out, and surveyed the damage, which was considerable. When I returned the car, much to my surprise, the rental company was not really bothered. This was evidently a not too uncommon event.

19.03 CONVINCING THE USA TO COMPETE IN THE OLYMPIAD

On returning to the United States, both Arthur and I were very keen the USA should take part the following year. The decision was to be made at a meeting of the AIP (American Institute of Physics), and to our surprise, there was a strong feeling against participating. The principal objection was that ultimately (it turned out to be about ten years later) the USA would have to host the event, and since one of the participants was Cuba, we would have to invite them. To me, this seemed like a great idea, but others thought there would be an immense amount of political trouble. Anyway, we carried the day and went to the next Olympiad, which was at Harrow School in England. I had made the point that having it in an English speaking country would be to our advantage, rather than in the Ukraine, say.

19.04 THE PHYSICS OLYMPIAD IN ENGLAND

Once the decision had been made that we should participate in the English Olympiad, the big question arose, out of a country the size of the United States, how could one select the six top contestants to go to England? This is no easy task, with millions of possible competitors. Clearly, it would be impossible for us to set and grade millions of papers. In the end, we advertised everywhere we could, that we would send a test to any teacher who requested, and let them administer and grade the test to whichever students they thought viable. The questions were multiple choice. They would then send us the results. We in turn would pick the best one or two hundred students, get the teacher to administer a second test, and we would grade it. That way, we would select the best twenty students, and these we took to the University of Maryland for coaching. There was some fear that they might all turn out to be of Chinese extraction, or something like that, but in fact we had a very wide ethnic and general background – all the students were very bright. Two were girls, one of whom I met at Baltimore Washington International Airport. She was quite small (but very bright), and I had difficulty realizing she was our participant. We were put up in very pleasant dorms at the University of Maryland, and every day we had the students solve problems, and compare techniques. We had them compete in groups of four, demonstrating their solution on the chalk board. We also had them do laboratory work. A couple of graduate students helped us, and we had one or two very eminent speakers, such as Nobel prizewinner Val Fitch, to stimulate the students. It was not all work however. Ultimate Frisbee turned out to be the game of choice. At the end, the students had to be rated and six selected. However, we emphasized that we were a team of all the participants.

We visited the US Department of Education, and had our picture taken with the Minister for Education William J. Bennett. He walked in, and stood in the middle of us, ready for the shot. Then he said "Do you realize the intelligence quotient for this group has just fallen twenty points?" However much I disagreed with him on other matters, I could not get mad with him after that.

We also visited the National Academy of Sciences, and had our picture taken with the Einstein statue.

19.05 TRIP TO ENGLAND

We flew from Baltimore Washington Airport to England, arriving a day early to get acclimatized. We stayed overnight in a hotel in Kensington, taking a walk around Kensington gardens. The following day we took the underground to Harrow on the Hill. Harrow school was considerably smaller than I had realized, and the rooms were quite Spartan. A large board listed the previous occupants of the house I stayed in, among which were several Churchills, though not Winston. He had attended there. The leaders met to set the competition questions, which took quite a while, then they had to be translated into the languages of the participants. It took almost as long for us to translate from English to American, as it did for the Chinese, who wrote the whole thing out using brush strokes. We had thought no translation would be necessary, but Eisenkraft pointed out that in the US, we use the word "slope" rather than "gradient" for a curve, etc. It was essential to get this straight, because the students could not ask questions during the test. The following day, the participants took the test, while the leaders met to decide on changes to the regulations, and a number of other things – the following year's venue, etc. We had one day off, when we took a cruise on the Thames, and visited the Greenwich museum – marine exhibits, plus the old Royal Observatory. Then came the practical exam. We had a nice party at the end, giving one another small gifts. Then back to America. The students unfortunately had missed their graduation prom, so we had a party on the plane.

19.06 THE IYPT

While at a conference on simple experiments for students in Duisburg Germany, I was drinking beer with other participants when they mentioned the International Young Physicists Tournament (IYPT), which I had never heard of before. It turned out that this is not an examination like the Olympiad, but consists in solving what might be called research problems, seventeen of which are set on the World Wide Web in the fall of each year. The high school students do research on these until the following June, when they meet together to compete, each team presenting solutions to the problems posed, which are criticized by the other teams and graded by a set of jurors. I failed to interest the AAPT in sending a US team the following year, but by great good fortune, the North Carolina School of Science and Mathematics had a windfall in the form of a bequest from a former student, Joseph Britt, which made the trip possible. In May 1999 five students (three very lively boys, and two attractive girls), Hugh Haskell, Charles Britton, (teachers at the school) myself and my wife set off for Vienna. Gertrude and I arrived a couple of days early, and were put up at a very interesting pension, where you had to go up to the second floor (first, English style) to get in, shops occupying the ground floor. It had obviously been converted from a residence, and they had put in a curious shower in a corner of the room. It was, in fact, very romantic, reminding one of the heyday of the Austro-Hungarian Empire, high ceilings and tall windows, but it lacked the convenience of a Holiday Inn. Personally, I much prefer to experience this kind of living, to get the feel of what it was like in a bygone era, but one must put up with the problems that arise, which accompany it. No private toilet for example.

19.07 TRIP TO VIENNA

We took a day or two to see Vienna, and then moved out to the Catholic School where the competition was held, in the outskirts of Vienna, at a place called Strebersdorf, which was in the wine country surrounding Vienna. We were given tickets allowing us to travel on public transportation, primarily the "electrische" – tram in England, trolley car in America. These were articulated, so you could stand on a rotating platform in the middle, while the car moved around you. I loved this! Our accommodation was comfortable but simple – and rather hot, since, of course, there was no air conditioning, and it was midsummer. The food was good, and it was interesting to meet all the other teams. They were much more formal in both dress and attitude than our people.

Since Strebersdorf was in the wine region, walking around one saw all the old houses and vinyards. At some time in the dim and distant past, they devised the "buschenschank" idea. A former empress I believe. Although the house owners were not allowed to sell their own wine, once a year they could hang a bush outside their house, and anyone could come and sample the wine that day, to see how good it was. One day I was walking around, and heard the distant sound of a brass band. I followed the sound to a delightful house and garden, where the Strebersdorfer Dorfmusik was in full blast, a celebration with brass band, and a wine princess. I sampled umpteen different kinds of wine, and rated them. At the end of the afternoon, my wife and I were feeling quite mellow. This also represented a buschenschank for the Familie Stampfer. At the wine shops in the evening, you sat down with whoever was there – we met some interesting people that way.

After the teams were eliminated, the finals took place in a palace, complete with chandeliers and much baroque ornamentation. Germany won, and Russian Georgia was second. Dinner, provided by the governor of Vienna, was held at the city hall – the Rathaus, a very elegant affair. As is usual in Europe, the participants were provided with wine. Some of our students were a little chary of this, because of course, they were not allowed to drink in North Carolina.

After the end of world war two, the Austrians were devastated, and one Viennese child, Hanni Bodenseer, went to live with my wife's family in Copenhagen. This resulted in a long lasting relationship. She was not in Vienna while we were there, but her sister Alice Werany was, and she accompanied us on some of our expeditions – up the television tower for example. While we were in the subway station, she suggested we go up the Schneeberg, a mountain not too far away. This sudden thought led us to ask the subway ticket purveyor whether he could sell us a ticket. He said he had never done this before, but intrigued by the problem, he attempted, and after about five minutes, Gertrude and I had tickets. Alice then thought perhaps she would go too. She normally got a senior discount, and the ticket guy managed to get a discount ticket for her too. We felt a little guilty, because a queue built up behind us while he was doing this, but he was quite pleased to prove he had this ability.

The following day we set off by train, ending up at the small village at the foot of the Schneeberg. We then took the cog railway (Zahnradbahn) to the top. For some unfathomable reason this was called the "Salamander". The seats were at a steep forward angle as we started, but as the ascent proceeded, they compensated for the steep angle of the mountainside, so we sat level. Disembarking at the top, it was very cold, but a magnificent view, with lots of snow. We had a meal up there, then took the "Zahnradbahn" back down again. Waiting for the train back to Vienna, I got talking to Alice. It was a bit odd. I spoke English and German, but no Danish, Alice spoke Danish and German, but no English, and Gertrude spoke English and Danish but no German. It sounds like one of those puzzles, but we got on fine. Alice and I conversed in German, and Alice and Gertrude spoke Danish. We arrive back in Vienna in good shape, and subsequently returned to America.

19.08 DON FRANKLIN AND THE IYPT IN FINLAND

It appeared nobody was interested in sending a team the following year, but by dint of the exemplary efforts of Don Franklin, a teacher then at Battery Creek High School in Beaufort South Carolina, a team was put together for the 14th tournament in Finland. Unlike our first team, which, coming from a state wide science school, was quite brilliant, the second team was basically composed of ordinary everyday students. Nevertheless, they represented a very interesting collection of individuals, as did the two Finnish students set to look after us. Perhaps the most unusual student was Richie Parker, a black student with no arms, having been born this way, who was quite brilliant. It amazed me how he managed. In Beaufort he drove a car, using his feet only, and following him I was surprised how well he drove – better than most people. We had two girls and three boys. They were very lively, but because of their background, they did not do well against the best teams in the world. The first day we did not do too badly, but ultimately we came in last, which was not too surprising.

We were treated well by our hosts, but the food in the cafeteria at Espoo, Finland was a bit Spartan, if healthy, driving some of the students to patronize the inevitable McDonalds. The school was directly adjacent to the shopping center, similar in many respects to those in the United States, but one of the things we don't get here were street performers. A group of three, including a massive bass balalaika were playing classical music in the middle of the street.

We noted the U.S. team had special problems at the international tournament. It might be thought that since English is the competition language, the U.S. team would have an advantage. This was not the case. Even though the US teams spoke with a slow Southern drawl, the Europeans (both participants and jurors) still had difficulty understanding the dialect. We told our team to speak extra slowly and clearly. Similarly, the U.S. team had difficulty understanding the Europeans and their European-style English.

In the past, no British team participated, and I asked Cyril Isenberg, who ran the Olympiad, why not? He said that the research involved for the questions distracted the students from their conventional courses. My own feeling is that this is the advantage of the process, and I note recently Britain did participate, as did 28 or more other countries. The number of participants increases yearly, but we still have little support in the U.S.

19.09 INTERNATIONAL PHYSICS TEACHERS – LONG ISLAND UNIVERSITY AND UCLA

I was invited to participate in a conference or international workshop for physics teachers at Long Island University, South Hampton, organized by a mad Russian named Edward D Lozynsky in the summer of 1987. He is a remarkable man, together with his wife Tatiana (they are now in a film). My wife went too. The thing was not well organized, in spite of which, we all had a very good time, and I think the teachers learned a lot. I held a physics Olympiad one day, including the egg drop competition. How quickly can you get an egg to the ground without it breaking? All kinds of methods were employed. A parachute proved most popular. Some important people came. One was Nobel Prize winner Sheldon Glashow. It was nice to go swimming in the Atlantic, if a little chilly.

The following year the conference-workshop was held at UCLA. Again, this proved a most congenial environment. I held the Olympiad at Pauley Pavilion, where later that year the presidential debate was held.

20 UNITARIANS ETC.

20.1 Methodist & Church of England

My religious background is definitely peculiar. I was christened a Methodist. My maternal grandparents were Methodists – my grandfather worked at Walker's tannery, and the Walkers were all Methodists, and, just as in the past if the Duke of a village was Catholic, so were his subjects, the workers at Walker's tannery were Methodists. This fact lead to my inheriting a nineteenth century encyclopedia which has proved most useful. It was given to my grandfather for his contributions to the PSA brotherhood. This is not connected with the prostate test, as it is today, but was a "Pleasant Sunday Afternoon", a kind of adult Sunday school. I noted, when last in my home town, that the rather austere Methodist chapel they attended is now something quite different, a library or a shopping mall of some sort. My father's family were church of England – which basically meant nothing, since they never went to church. Anyway, since most of my friends were C of E , this is what I became, and I was confirmed in that faith. I used to go on walks with them after church, and we even had a couple of girls accompany our assembly of boys. At college, I never had time for church – apart from chapel which I rarely attended. In Australia, I went to church only very occasionally, the nearest being quite far away. I am sure Canberra has many more churches now, the population having jumped by a factor of ten. I got married in St. John's Church of England by coadjutor bishop Arthur.

In the USA, I started going to what was then the Episcopal church of the Trinity here in Columbia SC, which I found ultraconservative. It was segregated. It is now the cathedral for the diocese of northern South Carolina My wife had heard of this small group of people called "Unitarians", so we visited them. They were meeting in odd places – shops, members' drawing rooms, a masonic temple – you name it. We got ejected from the Masonic

Temple because we had one black member (of whom we were very proud – we had so few). Unitarians had similar religious ideas to mine – I had never felt that miracles, such as the Virgin Birth, etc. fitted with my scientific background. Nevertheless there is something very fishy about this universe. Why are there laws which govern it? The anthropic principle is one possibility. It suggests there may be a vast number of universes, but this is the only one in which humanity could ever have existed. Anyway, no conventional church seemed to have any attachment to such ideas, and Unitarians left it to the individual to make such decisions. Our fellowship started out being much more humanistic, but has since become more mystical "spiritual" they call it! But as the chronic card gambler said, when he was found in a game he knew to be crooked, "It's the only game in town".

Abroad in South Africa and also in Germany, I could not find a similar church. My colleague, Carl Rau, induced me to avoid paying Kirchensteuer (church tax) since I was not attending church, but the shenanigans necessary to avoid paying this tax are very demanding in Germany, so in the end I paid.

Back in the United States I was asked to run for president of the Unitarian Fellowship here in Columbia. Unbeknownst to me, or the slate I was on, a group of members were opposed to us, and voted us out, much to my chagrin and surprise. Most of the individuals who were voted in left after about six months, leaving us with a white elephant in the form of a lot of land, which we had to dispose of. You learn much from experience! Anyway, sometime later I was elected president, and had to sign for our new church building – a vacant old episcopal church.

20.2 BUYING A CHURCH

The archdeacon, Beasley, dealt with the sale, and he confided in me that he was the first student at the University of South Carolina ever to receive a degree in physics. Quite a coincidence. He said most of his work was done by reading books.

After we moved in, the Episcopalians affixed a sign to our door affirming our building had been deconsecrated. Made me think of bell book and candle. It was a delightful old church, and the air-conditioning had been affected by removing part of a window which had Christ's feet on it and replacing it with the air-conditioner. We called it "our savior of the air conditioner". The church had been erected as a mission to the nearby cotton mill, Olympia, round about the turn of the century, 1900. It required a considerable amount of fixing up, which we did ourselves. The pipe organ had several notes missing, and our organist (Molly Jones) had to use considerable ingenuity to avoid these notes when playing. We tried mending it, but an old electro-pneumatic machine is difficult to repair. Eventually, the instrument was sold. After several years there, we moved to our present fellowship, in an old synagogue at the corner of Woodrow and Heyward streets in Columbia. For quite a while we shared it with the Jewish congregation, since they met on Saturday, and we met on Sunday We have gone through many vicissitudes, but still survive.

20.3 CHRISTOPHER AND MICHAEL

Unitarianism is a very liberal religion, so what can a Unitarian son do to rebel against his father? Well, my son Christopher converted to Catholicism. In fact, I was not opposed to this, because it showed he took a stand and was interested. However, as with all converts, he wanted to become a priest, which did not appeal to me, not being a devout Catholic mother. We were returning from SUUSI, the Unitarian summer institute, one year, and stopped off in Charlottesville, where Christopher was a graduate student at the University of Virginia. We all went out ballroom dancing one night, Christopher bringing along a very attractive young lady. At dinner, Christopher said, "Oh by the way dad, I've decided not to become a priest". This was his way of telling me he had become engaged to the young lady, Karen, now his wife. There are some advantages to having a Catholic son, one of which is that I now have four delightful grandchildren.

Michael on the other hand appears to have developed a very similar religious persuasion to my own – he even worked for a while at a Unitarian Church in Boston!

21 SUUSI AND SWIM

21.01 SUUSI

SUUSI stands for Southeast Unitarian Universalist Summer Institute. I found it was a unique institution, particularly if you are living in the ultraconservative repressive Bible belt of the deep South. In fact it kept me going to find out there were people around who did not object to kissing on Sundays.

When I first heard of SUUSI, it was meeting the last week of July at Radford, a small college in Virginia. SUUSI had gone through many vicissitudes. For example. they got kicked out of an establishment in Fontana because the kids kept running the elevators up and down. I went with Sue Folk, who had been a SUUSI participant for several years, and later was director for a year. In those conservative days, SUUSI was like a breath of fresh air – it opened up the mind and the spirit.

Apart from the people there, and the "serendipity" (accidentally discovering something fortunate) each night, with the dancing and fancy dress, there were the workshops. I gave one on the "State of the Universe", which discussed everything that had happened in science over the year, and Sue and I ran a play reading workshop for many years. I always enjoyed that – I think most people wonder what it would be like to be a villain or a hero. I always felt the villain had the best of it – but in a reading one can be one or the other with no recriminations. We actually put on performances once or twice. However, the workshop with the most controversy was massage. This was led by Scotty McDiamid, a licensed massage expert from Florida. He demanded that the best massage had to be done in the nude. There was never anything overtly sexual about it, but people of course imagine the worst (or best!). Anyway, we had to move to another room and shade the windows – and even there the board fought us. The kids program was the best – with "dirty day" when they could get as muddy as they wanted to. The tubing over

McCoy falls was also a delight, and rafting down the New River, which was very exciting when the river was running high.

21.01 SWIM

I always felt very depressed between Christmas and new year in Columbia. The weather was dreary, the university was closed – even the library was shut. So, I went down to sunny Florida to a Unitarian get-together called SWIM – Southeastern Winter Institute in Miami. I believe the first time was sometime in the late seventies and we went the second time it had been held. A group went down from the Columbia Unitarian Fellowship, staying with a friend on the way down. There was little accommodation, so we all slept in sleeping bags in the Sunday School facilities of the Unitarian Church in Miami. It was like sardines. There were workshops and services and outings. We had meant to go sailing in a large sailboat, but it was out of action, having landed on a head of coral, so a bunch of smaller sailboats was rented, and I, probably mistakenly, was made skipper of one. We sailed out to Key Biscayne and had a nice time at a distant waterside park on the island. However, on the way back a thunderstorm struck. Luckily, my boat did not have a large sail, but we had problems – like peeing off the stern because there was no toilet. We arrived back after the others, who heaved a sigh of relief on our return.

21.02 THE NUDIST COLONY

We had a number of very interesting workshops at SWIM. One was a visit to a nudist colony. I don't know how they arranged it, but it was a very enlightening experience. Firstly, we all felt very embarrassed as we got undressed, but the strange thing was, as soon as we were all completely naked, embarrassment vanished, and we became very gregarious. Sex really never played an obvious part. We swam in the pool talked and so on. Nearby a crop duster plane passed over on its way from the landing strip to the fields. Each time it passed over it came nearer and nearer. Of course the pilot knew where the colony was, and as it passed right over, we waved, and he dipped his wings in salute – he obviously was watching!

21.03 LOST IN THE MOUNTAINS

Round about thanksgiving in 1980, I had to give a talk at Clemson University, and Gertrude and I decided it would be an excellent time to go camping, and doing a little hiking in the foothills of the Smokies. So we hied ourselves to Table Rock State Park, which is where the mountains start. It turned out Gertrude's daughter Karen was also there, on a cycling expedition. We erected our tent in the park, in nice weather, then started up the mountain the following day. We reached the level top of the mountain, and met Wade Batson, a well-known professor of biology – botany was his subject – with several of his students. He told us there was a terrific view from the rock at the end of the mountain – and so there was. (Batson was a well-known character, who led botanical expeditions which were very popular. When asked one time why he walked so fast ahead of his students, he ostensibly replied that it was so he could put his foot on specimens which were unknown to him!).

However, on the way back we wandered off the trail somehow, and ended up down near the reservoir there. We then realized we were lost. There was a rude cabin down by the water, and we spent the night there. In the dusk, Gertrude thought she saw weird figures in the fog, and I believed I could hear moaning from the top of the nearby mountain. The following day we persevered, trying to find a path out, without success. Finally on the third day we were rescued by a water works employee, who came up the lake in a boat and saw us. I had Gertrude making a raft out of lianas and logs at the time. Later that year, when we were in Denmark, Howard Boozer sent us a cutting about a student who died in more or less the same place where we had been, under similar circumstances, some two months later.

22 CAPITAL JURY

22.01 CAPITAL JURY

University faculty rarely get called for jury duty in important cases – because they enjoy controversy. So when I was called for duty I went down to the state court house with little trepidation. The case was the murder of one of our coeds, Barbara Rossi, a beautiful girl, who had been abducted from the parking lot of a nearby shopping mall (Woodhill) as she got into her car. She had been raped and shot dead in an isolated area a few miles away. It was an important case, because women felt it could happen to any one of them – in broad daylight at a shopping center. In the jury selection, I told the judge that I was a Unitarian, and we are basically opposed to the death penalty. The judge said, "If you felt the criminal was so bad, could you see your way to imposing the death penalty?" I said I felt it unlikely, but yes, I was not fundamentally opposed, although it would have to be an unusual case. Much to my surprise, I was selected. It was not until many years later I found out why. All three suspects were tried together, an event I had never heard of in England. The lawyers felt the degrees of guilt were vastly different between the defendants, and that university people would argue over this, and bring in different verdicts for the three (which we did). A more conventional jury might not bother to distinguish and send them all to the chair.

22.02 OTHER JURORS

It turned out that John Adams, a member of the music faculty at the University, and our most celebrated pianist, was also on the jury, as was a former student of mine, married to a purveyor of Persian Carpets. In order that we should not be influenced, we were sequestered for two weeks during the trial, at a motel on Main Street and near Elmwood, now torn down. It was an unusual experience, because until all the evidence was in, we were not allowed to discuss the trial – the result was that after dinner in the evenings we played Trivial Pursuits. Whenever I play Trivial Pursuits now, I think of that experience. Luckily, the jury got on very well with one another, and, although we differed on some points, we were in general agreement about the case. This is not always true. The bailiffs said the jury of the previous trial of this nature had not got along with one another at all, leading to much friction, and a very unpleasant situation. The foreman of our jury was an African American, who, as I recall, ran a gas station, and had been a prize fighter. He was rather quiet, but as the case progressed, he mentioned his daughter had been abducted and killed under somewhat similar circumstances. Nevertheless, I felt he was very open and fair about the case. We were unable to communicate with anyone from outside, and were looked after by the bailiffs. These were police officers, and we had some interesting tales from them. Apparently, it was usual to send the rookies to entrap prostitutes, who would not know them, of course. They were not particularly keen on this, because the girls were likely to hit them with their handbags when they found out, and the officers were somewhat sympathetic to these girls, who of course, were merely plying their profession, and not hurting anyone. Nevertheless, the locals would complain their neighborhood was getting a bad reputation, if nothing was done.

22.03 SEQUESTERED

Adams had a birthday one night, and we were taken out to a nice restaurant, where, by coincidence, some of his friends were dining, and who came over and wanted to wish him happy birthday. They were completely confused by the bailiffs, who told them even this would be in violation of the law.

22.04 SENTENCING BY JURY – DEATH

The discussion over the verdict was most interesting. There were two trials. The first was whether the defendants were guilty or not. This was easy. In South Carolina "The hand of one is the hand of all" – so all were guilty. The second part of the trial was the sentencing – this was much more difficult. One defendant was definitely mentally defective so he, and the second guy who had been dragged in to it, were given life. The third member was a really evil individual – after much discussion, he got the chair. However, it later turned out that the judge had made an error in our absence, and this one was retried, about nine months later, and in Charleston. He also got life – the difference was, between the two trials, the law had changed so that life really did mean life. I was not sorry – I am basically against capital punishment, but this guy did deserve more than the others. We were told, after the trial, that his girlfriend was present in the front row of the audience. He had previously tried to kill her, but the gun would not fire, so he had beaten her with the butt of the rifle. Her face was still badly disfigured, and bandaged. However, the rules of evidence prevented this tragic event from being brought up in the trial

My conclusion was, although the legal system creaks and groans in this country, in this case at least I felt justice had been done.

23 CANAL CRUISE

23.01 LEEDS AND LIVERPOOL CANAL

When I was about twelve years old, I used to cycle over the moors above my home in Bolton. Crossing the moors was the Leeds and Liverpool canal, an eighteenth century transport predecessor to the railways. At that time it was defunct, and not used, although the locks and other equipment were still serviceable. I always wanted to travel on this canal. After I left England, there was a great revival of interest in such canals, not to carry goods, but for recreation. Now they have become very popular. I arranged to book a canal boat for a week on the canal in Yorkshire, so after a visit to Denmark with the relatives, we flew from Billund to London. We left Libbie and Debbie (Gertrude's grand daughter and her mother) in London, and took the train to Skipton in Yorkshire. It took only two and a bit hours, whereas when I was a child it used to be more than half a day. Change at Leeds, and head for the dales. We were met by Ian, (who ran the canal boat company), at the railway station, and he took us to Pennine Cruisers, on the canal in the middle of Skipton, just where the branch canal to the castle starts. Ian showed us the basics of the canal boat – how the very slow Diesel engine starts, etc. (chug, chug, chug), and we spent the night on the boat which had "all mod cons" – toilet, fridge, stove etc. We had delicious fish & chips for dinner.

23.02 LOCKS AND BRIDGES

The following day, (Tuesday) we started off along the canal in the direction of Lancashire. Ian saw us through the first canal lock and swing bridge, to make sure we knew the tricks how to operate them. This proved not to be a simple procedure. Firstly, the lock mechanism had not changed since the 1700s. Everything is mechanical, and worked by hand (and butt!). If the lock is empty (low water) and you wish to go uphill, you run the boat into the lock and shut the lower gate. This is no mean feat. You go to the wooden arm which sticks out eight or ten feet from the top of the gate, and apply your butt to the outer end, thus forcing the gate to swing shut. The gate is very heavy, and requires considerable effort to shut it. When both gates are shut, they form a V pointing uphill. You make sure the paddles (shutters) (which open to let water in or out of the lock) are down, so no water leaks out, and go to the upper gates. You wind a huge crank, to raise the "ground paddles". These are forward of the gates, and allow water to run into the lock from beneath. This does not disturb the boat in the lock too much. When the lock is half full, you open the gate paddles. Water runs through the gate, producing a large disturbance in the lock, which then fills. When the level inside and outside the lock are the same, you open the top gates (with much effort). I had thought, at my age, this would prove an insuperable difficulty. Not so — you have to apply a continuous force on the gate lever, but it moves, slowly but steadily. Strangely enough, I started the cruise with backache, but this had gone when we finished.

When the gates are open, the boat moves out. This part was done by Gertrude who had remained on board to work the motor. I then shut all gates and paddles, and re-boarded. The swing bridges were also a problem There were a lot of these below Skipton. Ostensibly, one member of the crew gets off the boat, crosses the bridge and unlocks it with the British Waterways "handcuff" key, then spins it open by applying his or her butt to the lever arm. Again, much force is required, but the pivot is well lubricated and it works. The boat (piloted again by Gertrude) passes through, the butt is applied to rotate the bridge in the opposite direction, one crosses the bridge and gets on the boat. It took me a while to realize why the bridge operating system is on the wrong side of the canal – this forces you to close the bridge in order to cross and board the boat! Some bridges had an electronic system which made lights flash, lowered a boom on the road and rang a bell before the bridge opened. Such bridges were supported at each end as well as the middle.

After we had been through a lock, and bridge, Ian left and we continued on. We were assisted locking through by the British Waterways lock keeper at Gargrave. He also gave us some good advice, such as that Gargrave was the last useful village until we were over the crest of the pennines – which proved to be true. It was then 3.30 pm, so we called it a day, moored with the other boats against the tow path, I went into the village and Gertrude took a nap. Mooring consisted of knocking two steel pegs about two feet long into the canal bank, and attaching ropes fore and aft. Further advice from the lock keeper was that there were two good pubs in Gargrave – the Old Swan, and the Mason's Arms. The latter had more character, but the former was nearer, so I hied off there for a quick pint. Boy that beer tastes good on a hot day. As I recall, we ate bangers (sausages) on the boat.

Following morning we set out. At one point I walked ahead to see how the people in front were doing. Turned out it was a boat with a man and his wife, a ten year old boy and a baby in a pusher on deck. The lock was empty, but the dad had just fallen in the water – I did not see how. The water in the empty lock was up to his chest, so he was able to walk over and climb the ladder which is in the wall of every lock. The baby, left alone on the boat, was not too happy. Anyway, after the lock was filled it was possible to get back on the boat with no trouble. The boy confided he had fallen in last time.

23.03 FOULRIDGE TUNNEL AND THE COW

On we progressed, and were caught up with by another boat. The guy in this pointed out we would get along quicker if we entered the lock together. This we did, the one following the other but lying parallel in the lock. In filling and emptying the lock, he took one side, and I the other. It was also safer for the boats, which were tight together, and could not move. We did this for several locks, then he went on ahead. We stopped for lunch. later going on and tying up at the Anchor in Salterforth which is just over the border in Lancashire, and is an ancient pub, with stalactites in the basement. We had dinner, a pint of "bitter" for me, and "mild" for Gertrude. The following morning we went on to Foulridge tunnel, 1640 yards long built well before 1820. Since the tunnel is only wide enough for one boat, you have to wait at the end for a light to turn green, so your boat can enter.

Successive boats must leave a wide space between them. I got through without bumping once. I found the way to steer is to look at the pattern the one headlight made on the wall, and make sure it was even on both sides. We went on to the top lock at Barnoldswick. We had thus travelled the maximum distance at the highest level of the canal. We turned and went back – I did not feel attracted by operating more locks than I had to. The light was green over the tunnel, so we went straight in – you could see the daylight at the other end, growing larger and larger as we progressed. We stopped at a restaurant on the dock at Foulridge, where a problem arose. Gertrude only had Travellers Cheques, and I had run out of English money. The proprietor of the restaurant was very helpful, and drove us to the post office to find the rate of exchange (I had plenty of dollars), so we got to see the very pretty village. He then exchanged 40 dollars to pounds, to pay our bill. Since the twin towers incident, and foot and mouth disease, they had not had as many Americans. The restaurant had very good food. I had delicious lamb, and Gertrude had a delectable mushroom soup. This was where the "cow in the canal" incident occurred. A cow fell in at the far end of the tunnel, and somehow struggled through, being revived near a pub (now called "The hole in the wall") by an alcoholic beverage.

We continued on, and went through lock number 168A, tying up for the night. However, this is just below a railway bridge, and we thought the noise of the trains might disturb us, so we went on a bit further. We had dinner, and went to sleep. I awoke about two AM to find the boat tilted at an angle of 10 degrees or so. I was appalled. Either the boat had sprung a leak, or the level of the canal had gone down. I could not imagine the latter, but the former seemed unlikely too. I was dog tired, and went off to sleep again. Waking in the early morning, the boat was now tilting at 15 to 20 degrees, which made it very difficult to move about inside. What to do? My first thought was to call the owner, Ian. This, however, would mean walking miles to find a phone. Upon going outside, I found the bank, made of wood, was wet to about a foot above the waterline. It was thus clear that the level of the canal had dropped a foot or more during the night. That the canal could leak that much in such a short time seemed impossible. The only solution was that there was a lock downstream, out of sight, where this had happened. As an experimental physicist, it seemed clear that if the level had gone down this much through the downstream lock, we ought to be able to restore the level through the upstream lock. It seemed strange to me that the general level of the canal could be affected so much by water through the locks. Anyway, I went to the upper lock, and filled and emptied it. Sure enough, the canal level went up over an inch. Flushing the lock ten more times and we were able to float off. I am still baffled by this, because the lower lock had a weir and stream bypassing the lock, which, one would have thought, would keep the canal level steady. Anyway, we heaved a great sigh of relief, and went on.

We tied up at Gargrave again – delightful place. Gertrude had seen that they were holding a tea dance in the village hall there at two in the afternoon, and wondered what this was. In my home town Bolton such an event generally involved tea and cakes, accompanied by three old ladies who played violin, cello and piano, to waltzes, foxtrots, and quicksteps. We got to the hall about three thirty to find a group of people dancing to a record player. It was a rather complicated couple dance, and each pair seemed to be doing exactly the same thing with the most gloomy serious expressions imaginable. It was clear this was not our thing, so on we went, and rested at the stepping stones on the Aire river (remember an airedale dog? This was the dale where they come from) until the pubs opened.

Several kids were playing about in the water. It was a beautiful day, but the water was not exactly warm. We walked on to a very traditional sweet shop. They had sweets (candy) in jars which I had not seen in thirty years – including "mint imperials" (Grandma Edge used to love these), dolly mixtures, and Hartsfoot rock – (rock is hard candy – remember Brighton Rock?) good if you have a sore throat. We went back to the Swan, and sat in the beer garden for a pint – then back to the boat for dinner. The following day, we returned to Skipton but did not stop. It being Saturday, it was crowded. We went on, following a boat carrying passengers on a short tour.

They came up to a swing bridge which opened to let them through. Then it started to close, although etiquette would demand that you let a following boat, namely us, through first. However, the bridge shortly opened again and let us through. The people on the bridge apologized saying that the propeller of the boat ahead had caught a chain, causing the bridge to close automatically. They had opened it again. Never a dull moment, even at 4 mph.

23.04 TWO MILES AN HOUR

Four and a half miles per hour was our maximum speed produced by the slow diesel engine, but on passing moored boats you are supposed to slow to two miles per hour, to reduce the wake. I did not realize this until I passed a mooring, and a girl popped her head out of the boat window saying heartily "slow down, slow down – I'm bouncing around like a pea on a drum!" I would start the engine in the morning, with some difficulty, and not stop it till evening. The stretch of canal above Skipton had many locks, but that below had no locks, but lots of swing bridges, of all kinds. We decided to stop at Kildwick, and tied up just below the swing bridge. We had an excellent dinner at the White Lion, a pub near the canal. There are always lots of pubs near the canal – for obvious reasons! Most of these date from the eighteenth century, when the canal was built, because travel on the canal was slow, which gave one plenty of time for refreshment on the way. We sat outside and relaxed in the gorgeous weather. A walk through the village takes you along a narrow road which crossed under the canal. It was very drippy. The following morning we turned back. It tried to rain, (without much success) the only time on our trip. Tying up at the dock in Skipton, we viewed the town. I went round the castle, and later listened to them try for a peal of Cambridge Surprise Major on the bells of the church near the castle. We had an excellent dinner at the "Woolly Sheep" on Main Street.

23.05 LONDON AND RETURN

The following morning, Monday, we had breakfast, packed, and Ian saw us to the train, which we took first to Leeds and then London, where I went to the hotel booking agent on the station at Kings Cross. He booked us for a very nice hotel, the Euston Plaza, where we spent the night, exhausted. We had dinner at a nearby Greek restaurant which had the most delicious lamb, and the most awful soup. Next day I took Gertrude to Harrods, which proved incredibly expensive, then the new Tate Gallery on the South Bank – what was the old Battersea power station. I was frankly not impressed, but they did have a wheelchair for Gertrude. What used to be the turbine hall is huge. In the afternoon we queued up for the London Eye. There was an incipient rainstorm which did not eventuate. The view over London from the pod of the wheel of the eye was well worth the wait. Then back to the hotel, and dinner on Euston station, which was the nearest place. I had a Burger King chicken sandwich – Gertrude had French pastry. The following day, taxi to Kings Cross, Thames Link to Gatwick, and home. Gertrude's daughter Karen met us in Columbia.

ABOUT THE AUTHOR

Dr. Edge was born in the dreary Lancashire mill town of Bolton (since reborn) and won a scholarship to Bolton school, an eminent public (and therefore private) school. He won a major scholarship to Queens' college Cambridge, where he obtained a triple first class honors degree in physics, and went on to obtain a PhD in nuclear physics. He emigrated to Australia to build an electron synchrotron (a nuclear accelerator) at the then new Australian National University in Canberra. From there he travelled to the University of South Carolina and has since taught and performed research at Stanford, Yale, Munich, Sussex and Witwatersrand Universities, and Oak Ridge, Brookhaven, Los Alamos, and Argonne National Laboratories. He was elected President of the American Association of Physics Teachers in 1997 and was examiner for the Graduate Record Examination. He was a leader for the first participation of the USA in the International Physics Olympiad in1986 and the first participation of the USA in the International Young Physicists Tournament in 1999.

Dr. Edge is currently a Distinguished Professor Emeritus in the Department of Physics and Astronomy at the University of S.C. He has numerous publications in nuclear and solid state physics and is author of a best-selling American Physics Teacher book "String and Sticky-Tape Experiments".

Made in the USA
Charleston, SC
15 October 2016